D1603576

Sonnets & Sunspots:

Sonnets & Sunspots: "Dr. Research" Baxter and the Bell Science Films

by Eric Niderost

BearManor Media
2014

Sonnets to Sunspots: Dr. Research Baxter and the Bell Science Films

For information, address:

BearManor Media
P. O. Box 71426
Albany, GA 31708

bearmanormedia.com

Typesetting and layout by John Teehan

Cover photo courtesy Wesleyan Cinema Archives

Published in the USA by BearManor Media

ISBN—1-59393-582-X
978-1-59393-582-5

Table of Contents

Acknowledgments

THEY SAY THAT WRITING is a solitary profession, and to an extent this is true. But no book can be completed without the help of others. This work is no exception. At the risk of sounding like an Oscar winner, there are many people who I'd like to acknowledge and thank.

The Baxter family: First and foremost Lydia Morris Baxter, Frank Baxter's daughter, was very generous with her time and her memories. She was very patient in answering my numerous questions—questions that in many cases only she could answer. Lisa Baxter Garber, grand-niece of Dr. Baxter, has not only been a great source of information, but a good friend too. Allison Blackburn (Dr. Baxter's daughter-in-law) was very generous in sharing her memories and letters. Dr. Baxter's other daughter-in-law, Kaye Baxter was also helpful. Karl Greenwood supplied me with much-needed information over the months.

University and Library Staff: USC Librarian Claude Zachary and USC publicist Allison Engel gave me great assistance over the last couple of years. And Archivist Aaron Spelbring gave me copies of a treasure trove of letters that Dr. Baxter wrote to his mentor Harold Colton from the 1920s to the early 1960s. James Blinn, CG artist extraordinaire, also gave me some interesting information. Head Archivist Joan Miller of the Wesleyan Cinema Archives was most helpful in researching and obtaining Bell Science film stills from their collection.

Dr. Baxter's former students gave me a good idea of how he taught in the classroom. These include Betty Frey, Pastor David R. Brown, Joseph Jares, Alan Simon, Meredith Churchill, William Owen, Theodore Cordes, S. Howard Wallis and Florence Nishida.

Other friends have also made wonderful contributions. Danny Kuhn, a new and very talented novelist, has given me great help and encouragement. Vincent Summers, a scientist who was inspired by Dr. Baxter, also helped with the professor's family tree research.

Special mention has to be given to three talented individuals and friends: Tom Sito, June Foray, and Louis Kraft. Tom and I go back almost twenty years. He's a very talented animator who had worked for Disney and other studios over the years. We share a love of history, and in recent years Tom has become a very talented writer and historian. He's now a Professor of the Practice of Cinematic Arts at USC, Dr. Baxter's old university.

June Foray is one of those few people who deserve the title "legendary." Though she's perhaps best known as the voice of Rocky the Flying Squirrel from Rocky and Bullwinkle fame, she had done scores of productions in the last sixty years or so. Louis Kraft is a very impressive writer and chronicler of western history—and a good actor, to boot! His encouragement and friendship has meant a lot to me over the years.

I also want to give a very special thank you to James Burke: a science historian who's perhaps best known for his *Connections* series. James took the time from his very busy schedule to not only read my manuscript, but also write a foreword. He was also very complimentary about the book itself. I am very grateful to him, and to all.

Foreword

This book is timely. It tracks and describes in intriguing detail what may be the first and last attempts, over a fifty-year period (by those of us in the mass media), to bring the general public to a better understanding of science and technology and the ways in which they shape our lives. From the first hesitant steps taken by the early 1950s university-station ETV producers to the fast-moving computer-generated images of today's PBS shows, this book reveals how science programming has tried every means to make arcane gobbledygook informative and fun.

Early seminal work was pioneered by men like presenter Frank Baxter and director Frank Capra, who got their stories across to the audience by means of cartoon characters, chalk and talk exposition, inspirational music, puppets, and (sometimes) earnest, talking-head scientists. Some of those first shows now look old-fashioned and amateurish, but at the time, any attempt to deal with science was cutting-edge stuff.

Over the decades, as science and technology loomed larger in people's lives (especially in the form of the Big Science of the Cold War years) TV began to reflect the growing public awareness that what was happening in science labs could make your day, or bring an end to life on the planet.

It became increasingly important to explain the relevance of science and technology and, above all, how it operated. So as the scientific disciplines proliferated and fragmented into ever-more specialist areas of activity, more recent audiences were treated to ever-more detailed investigations of such life-long geek pursuits as the neuronal structure of maggot brains, or event-horizons in the vicinity of light-year-distant black holes.

However, even in that last bastion of public service broadcasting, the BBC, it grew more and more difficult to justify budgets for low-viewer science shows when cheap cookery or house-improvement programs would bring millions to the channel. It began to feel as if the early dream of television as an educative medium was indeed no more than a dream.

But what has brought the question of mass media coverage of science and technology into stark relief (in particular because of the relentless commercial pressure to ignore serious subject-matter in the quest for bigger audiences) is the advent of the internet and the sudden, exponential growth of information technology. Even scientists are finding themselves and their subjects of study increasingly computerized. Algorithms now explore the unknown where once only the dedicated human researcher would venture. As for the general public, search engines for dummies give the impression that knowledge is freely and understandably accessible to all. Why watch science on TV (if you can find it there at all) when you can get all the information you want at the click of a mouse or the swipe of a finger?

But the real question is whether or not the public any longer needs to 'understand' science and technology at all. In the next few decades nanotechnology will build ubiquitous interactive networks of trillions of dust-mote-sized computers, embedded in every object (perhaps including people) and capable of storing and managing the entire corpus of human knowledge. These 'motes' will use big data and predictive analytics to scan the massive 'data exhaust,' that society generates with our every action, in order to identify what people need and want. And then innovate to suit.

In a sense, science will become 'invisible,' to us in the same way that we are not aware of the internal workings of our present push-button machines. Today we don't need to understand how a fail-safe refrigerator works to chill our food. Tomorrow the same will be true of genetic engineering or nuclear-fusion power-plants. We will leave such matters to the motes.

If this is indeed what lies ahead, then part of the reason we will be ready for it will have been those early television efforts by Baxter and Capra and the others to introduce science to the general public and help to make it an embedded part of the general culture. We have much to thank them for. This fascinating and meticulously researched book explains why.

– James Burke,
Menton, France

Introduction:
Reflections of a Baby Boomer

One fall day in 1961, Mr. Leo Bachle's science class at Calaroga Junior High School settled down to watch a 16mm movie. We waited patiently but with a growing sense of anticipation as Mr. Bachle skillfully threaded the long perforated strip of celluloid through the Bell & Howell projector. Films were always welcomed because they broke up the dull routine of schoolwork and transported us to places beyond the narrow confines of our Hayward, California, classroom.

When the movie began, the swelling strains of Ludwig van Beethoven's "Eroica" filled the classroom, and the title sequence announcing the subject was *Hemo the Magnificent*: the story of blood and circulation. The subject was new, but the series and format were not. *Hemo* was part of an ongoing popular science series bankrolled by American Telephone and Telegraph, the parent company of Bell Telephone. An entertaining mix of live action, science film clips, animation, and puppets, these films taught complex subjects in a thoroughly engaging way.

Hemo starred actor Richard Carlson and Dr. Frank Baxter, with the latter as a kind of generic scientist called "Dr. Research." Carlson had his moments, but to students Dr. Baxter was the real star of the show. He was warm, witty, and charismatic, with just the right amount of mature authority to make you believe whatever he said. Baxter was like a beloved older relative, in the mold of our other TV and film "uncle," Walt Disney.

Another science hero of mine was Dr. Earl S. Herald, Curator of Marine Biology at the California Academy of Sciences in San Francisco. He was host of the local show *Science in Action*, and I probably saw almost every episode because I was a Bay Area kid. The shows were interesting, and once, when I was about ten, I had the thrill of actually catching a glimpse of Dr. Herald walking by at the Steinhart Aquarium. But when all was said and done Dr. Frank Baxter was "*the* man," the supreme teaching hero, even role model, (I became a history teacher) of my world.

The irony is that Baxter wasn't a scientist at all but a professor of English Literature at the University of Southern California. From 1953 to 1970 he reigned supreme as the academic who taught a wide variety of subjects on television and film. Baxter was a pioneer of popularization, using the new medium of television to educate and entertain not just children but the public at large.

Frank Baxter was nothing short of a phenomenon. He proved that academics and their subjects need not be boring, and in the process helped resuscitate educational television—an infant that seemed already on the verge of extinction. His premiere series, *Shakespeare on TV*, took the country by storm, captivating viewers from nine to ninety. As his programs aired, there were library "runs" on books he suggested.

Dr. Baxter is an enormously important figure in two respects. First, he helped save educational television at a time when it was brand new and struggling just to survive. Operating on a shoestring, anemic, and wasted from lack of funding, public television was on life support in those early years. Baxter's programs gave it a much needed transfusion of interest, enthusiasm, and popularity that helped sustain it during this critical period.

Secondly, Baxter joined with Hollywood director Frank Capra to revolutionize the science and technology documentary. There were nine films in all, though Capra only produced the first four. These programs were first aired on television, but enjoyed a kind of immortality when they were transferred to 16mm prints and distributed for free in schools throughout the United States and Canada.

In 1961, it was estimated that in the years since the first entry, *Our Mr. Sun*, 76,000 American students saw one or another of these films each day, and an additional 15,000 a day viewed them in Canada. Over the thirty-odd years that they were in vogue, perhaps 200 million students have seen them, an astronomical figure by any definition. Many baby boomers were so inspired by Baxter and the Bell films they became scientists.

Baxter and the Bell films provided the foundation of the popular educational film, but its evolution was not yet complete. The British Broadcasting Corporation (BBC) and a science reporter named James Burke furthered the art by adding widespread location filming and historical reenactments into the "mix." He recognized, independently, something that Baxter and Capra had discovered years before: humor, models, and a sense of fun makes even seemingly "esoteric" subjects accessible. *Connec-*

tions was a major global success, validating again the techniques used by Frank Baxter and Frank Capra.

The television educational series reached a new maturity around 1980—the year when the ten-part series *Cosmos: A Personal Journey* aired on PBS. Presented by astrophysicist Carl Sagan, the series included many elements of the Bell Science "template:" a charismatic on-screen personality, brilliant visuals, and multi-topic episodes. But *Cosmos* added CG (Computer Generated) visuals, something that was unprecedented up to that time.

The CG work in *Cosmos* was partly done by James Blinn. *Cosmos* is, in many respects, the lineal descendant of the old Bell series. It's no coincidence that Blinn vividly remembers the Bell films he saw as a child, and was strongly influenced by them. He is a fan of the good doctor and his film work.

Popular shows like *Connections, Cosmos,* and *The Day the Universe Changed* followed the trail that Baxter and Bell had blazed thirty years before. The popularity of these shows and many others attracted viewers to PBS and helped it spark a new interest among the general viewing public. As these pages detail, educational TV owes a lot to that moon-faced, smiling professor of English Literature.

1
Origins of an Icon

In the last decade of the nineteenth century, America was like a newly emerging butterfly, struggling to escape the chrysalis of an agrarian, traditional past. The era used to be viewed in an almost nostalgic vein where men wearing bowler hats and sporting handlebar mustaches courted wasp-waisted "Gibson" girls. But Mark Twain, who knew the time well, called it the "gilded age." It was indeed a time where a patina of superficial glamour hid a base metal of corruption, poverty, and greed. Yet there was progress too, and the seeds of the modern world were being planted that would spout, for good or for ill, in the coming years.

On May 4, 1896, a child was born who was going to chronicle, and even explain, the wonders of this emerging brave new world. Francis (Frank) Condie Baxter made his "debut" in Newbold, West Depford Township, Gloucester County, New Jersey. His father, Frank Sr., was a salesman; his mother was Lillian Douglass Murdoch. The couple married in Atlantic City, New Jersey, in 1893, and Francis—Frank Jr.—was the only child born of this union.

The timing of young Frank's birth could not have been more propitious. Milestones of science and technology were being created around this time—the building blocks of today's world. Sixty years into the future, they would provide Baxter with ample material for his film and television programs. On June 4, 1896, just a month after Baxter was born, a young Henry Ford test drove his "quadricycle"—his first automobile—on the streets of Detroit. Thomas Edison met Ford in August of that year and insisted that he stick with a gasoline powered engine, not an electric one.

In 1958, Baxter would warn of car exhaust pollution and global warming in *Meteora: the Unchained Goddess.*

In 1895, Guglielmo Marconi demonstrated wireless telegraphy, the sending of messages through space. Frank Baxter would discuss the effects of solar flares on radio communications in his premiere Bell Science effort, *Our Mr. Sun* (1956). But perhaps the most significant event occurred just eleven days before Baxter came into the world.

On April 23, 1896, the Vitascope movie projector had its first theatrical exhibition at Koster and Bial's Music Hall in New York City. The audience didn't know exactly what to expect, because they found themselves sitting in front of a twenty-foot-square white screen. When the house lights dimmed, the screen suddenly came to colorful life, a veritable feast for the senses that left patrons transfixed by the images and awed by the sheer wonder of it all. A bevy of dancing girls cavorted around, a couple of comedians did a comic boxing match, and waves curled and foamed and broke upon a sandy beach.

This was in a very real sense the birth of the motion picture industry, and though it was designed for mass entertainment, Baxter was going to ably demonstrate it could inform and enlighten too. Frank Baxter and the medium that he would do so much to transform were born about the same time.

Newbold was a tiny community that was small in area and population but rich in history. To the north was Big Timber Creek, with a meandering stream that finally emptied its water into the mighty Delaware River. The Lenni-Lenape Indians camped there, the same tribe that negotiated with William Penn. In 1624, the Dutch built an important outpost, Fort Nassau. Its exact location remains a mystery, but was probably located within Newbold's boundaries.

An old cattle track once wound its way through the hamlet, where bellowing beeves, protesting every inch of the way, were herded towards George Washington's hungry Continental Army. In the years before the Civil War, the house of a certain Thomas West was said to serve as a "station" in the "Underground Railroad," its mission to help escaped slaves on their way to freedom.

Newbold's westernmost tip touched the Delaware River, on New Jersey's western shore. The little borough was only about nine miles south of Philadelphia. The City of Brotherly Love was a natural magnet for those wanting employment, new opportunities, or simply the excitement of a bustling, dynamic urban environment.

Little seems to be known of baby Frank's formative years. The 1900 census shows the family still in Newbold, and presumably young Frank played, learned, and grew like any other child. But sometime during the next few years—possibly around 1904—Frank Senior left the family for good. This is not the first time that Frank Sr. had done such a thing. Before he met Frank's mother Lillian he had been married to a Lillie Selheimer, but the union ended when she died in 1886. Widower or not, Lisa Baxter Garber, Frank Sr.'s great granddaughter, says he had apparently left the first family even before his wife's demise.

The future Dr. Frank Baxter had a half-brother from his father's previous marriage, William Selheimer Baxter, but apparently there was little or no contact between the two men in later life. Lisa Garber says William and Dr. Frank look very similar, with moon face and bald dome.

The breakup of the Baxter marriage was going to have a profound and immediate impact on the future professor's life. Though the evidence is circumstantial, it seems as though Frank and his mother were cast adrift with little or no support from an absentee husband/father.

In the short term, that meant that Frank went to work at the tender age of eight. Child labor was common enough of the period; in 1900 there were 790,000 working children between the ages of ten and thirteen. But the Baxters were middle class, not working class or new immigrants. It seems that young Frank was sent to work simply because he had to help support the family.

The breakup of the family and his being sent to work was going to take an emotional toll as well. It certainly affected his interpersonal relationships for years to come. As a result of the separation, Frank's mother became a dominant—and eventually, domineering—influence on Baxter's life. Secondly, he found it hard to relate to young children in later years, particularly his own offspring.

Most mother-child bonds are strong, but the absence of a father made the link between Lillian and her son particularly durable. The bonds of affection provided emotional support in good times and bad, but when he matured, those bonds became fetters that prevented him from becoming fully independent until he was well into his thirties. Lydia Morris Baxter, the professor's daughter, says: "My father had a problem. He couldn't do anything without his mother. When he was young and invited out he would ask if his mother could come too."

Young Frank's first job was a water boy, serving glasses of water to patrons in box seats at Philadelphia's Civic Opera House. Another version, slightly different, has the eight-year-old handing glasses of water, backstage, to parched divas between arias. According to family tradition, Frank saw President Theodore Roosevelt give a speech, but the great man directed some of the remarks to *him* so forcefully the lad was scared to death.

In any event, Frank did find time to go to school, but the experience was less than edifying. One of the ironies of Frank Baxter's life is that the man who would become one of America's greatest teachers seems to have hated school. We know little of his actual academic record, but random comments he made over the years were anything but positive. Usually the remarks were made in the form of a joke because Baxter was blessed with a wonderful sense of humor.

In the 1950s, he recalled a history teacher whose teeth resembled walrus tusks. Writing in a 1956 issue of *Reader's Digest*, he remembered: "The little I found out about Mexicans came from memorizing long pages about the Mexican War. We were in the seventh grade (around 1908) and our teacher was a very tall lady with two of the most amazing incisors this side of the Arctic Circle. Those incisors bit much more sharply into my mind than the Mexican War did. I believe we finally won, but I am not sure."

On another occasion, while giving a speech at Brigham Young University commemorating the 400th anniversary of Shakespeare's birth, Baxter recalled learning about the Bard in the early years of the twentieth century. "I can remember back," Baxter explained, "to about 1908....I swear that this is so: we read a Shakespeare play word by word, and we parsed it. Word by word we dissected Shakespeare: we took him down to the very molecular basis of his language, and the ordeal went on for weeks."

Baxter continued, "Nothing came through to us of the drama, the humanity, the fun, or the poetry—but, by George we *had* Shakespeare. It was years before I recovered from this experience." Perhaps the memory of those torturous days actually made Baxter a superior teacher in years to come. He would strike out on his own, pioneer new ways of presenting material that could be entertaining as well as memorable and relevant.

Perhaps the most startling fact about Baxter's disenchantment with education is that he was a high school dropout. The man who earned his Ph.D. in Cambridge, and developed into one of the most honored educators in American history, left high school at the end of his freshman year to pursue fulltime employment.

The year was 1912. The Titanic set sail on its fatal voyage, and the country was absorbed in a presidential campaign that had Teddy Roosevelt stage a third party bid to win the White House. Sixteen-year-old

Frank Baxter landed a position as office boy on the Pennsylvania Salt Manufacturing Company. He stayed with the firm until 1917, rising to the position of accountant.

We may never know if he would have stayed an accountant the rest of his life, but in April, 1917, the United States went to war against the Kaiser's Germany. German submarines had been sinking American shipping, and President Woodrow Wilson declared that a war—a war to end all wars—would be necessary to make the world safe for democracy. Frank C. Baxter, twenty-one years old, duly registered for the military draft on June 5, 1917.

And so Baxter became a doughboy, part of the American Expeditionary Force that was sent to France under General John Pershing in 1917-18. Baxter was a medical corpsman stationed near Nantes, treating the wounded from the trenches. The Allied propaganda machine was in full swing, characterizing the Germans as beasts and "Huns." Baxter once commented that when some German POWS were brought in, he and his comrades half expected them to have horns and forked tails.

Baxter was stationed a long way from the front, but as a corpsman he no doubt saw the horrific wounds and suffering that war can produce. He seemed to like his work, and briefly—very briefly—flirted with the idea of becoming a doctor. Sometimes the supply line broke down. At one point his unit had to subsist on bread and bacon—and bread fried in bacon.

Doughboy Baxter emerged from the war unscathed—save for a case of salmon that fell on his foot. Recalling the incident with humor, the professor declared that the injury "gives me a picturesque limp on rainy days." On November 11, 1918—the eleventh hour of the eleventh day of the eleventh month—the guns fell silent. The Armistice was signed, and the Germans surrendered. Baxter was demobilized and went back to Philadelphia, where he took up residence with his mother Lillian.

Not long after his return, Baxter found himself at a crossroads—literally and figuratively. According to a story he told his students decades later, he was standing on a street corner in Philadelphia pondering what to do with the rest of his life. Conventional wisdom would dictate that he would resume accounting with the salt company, but Baxter had other ideas. The seeds of a radical new plan had been germinating in his brain for a long time, and now they started to sprout. Why not become a professor?

This was a perilous step—nothing in his background, or academic record, suggested that he would excel in the teaching profession. And he was all too aware he was a high school dropout. The issue would be decided by a coin toss. One side would represent the salt company, the other the University of Pennsylvania. Baxter flipped the coin into the air, the metal circle tumbling and somersaulting in a wide arc. The University won the toss.

Perhaps Baxter gave an extra push to the university side, because once he commented that: "I decided to do what I had always really wanted to do. I went into teaching because I thought it was the most exciting thing in the world." But he still hadn't fully decided on what direction to take, or what field to specialize in.

In later years, much was made of the fact that Dr. Research, his film alter-ego, was a scientist while Baxter was really a professor of English literature. But, in his college years, Baxter was indeed a scientist of sorts, with interests in zoology and archeology. He fell under the influence of Dr. Harold Colton, professor of Zoology at the University of Pennsylvania. Colton was interested in Southwest Indian culture, particularly that of the Pueblo Indians.

Baxter became his assistant both on campus and on the several archeological "digs" that the professor conducted in Arizona's Painted Desert. Serious Native American archaeology was in its infancy, and though Colton was far from a pot hunter, he admitted later: "we did not know what we were doing." One site they thought was a kiva turned out to be a pit house.

But these Arizona expeditions instilled a lifelong love of the desert. Baxter became something of an expert in desert flora, fauna, and geology, and Colton became a lifelong friend and mentor. In fact, Baxter's enthusiastic letters to the older man are a primary source of information during this period.

Young Baxter would have a wide variety of duties when the Colton expeditions were in the field. Besides assisting the "digs," he would drive, do camp chores, or provide various "gofer" services. When Baxter was invited to go to Arizona—apparently it wasn't an automatic thing—his enthusiasm was overflowing. "The joy of the Baxter family is boundless!!" he enthuses in one missive. "I do not know how to thank you! I shall chop, labor, laugh, dig…" Baxter's mother was also going to go, with Frank assuring Colton that she was "ten years younger" just in the anticipation of the trip.

When a party of geology students from Princeton visited Arizona, Frank Baxter joined Professor Colton in guiding the newcomers through the region. Eventually Colton and Baxter collaborated on a guidebook entitled *Days in the Painted Desert and the San Francisco Mountains.*

The young scholar also started to teach zoology, though he still had not earned his Bachelor of Arts. Baxter became interested in drama, and performed in several college plays. Something of a frustrated actor, he once complained, tongue firmly in cheek, that he was always cast as a "lawyer or doctor or in some nonconsequential role."

After graduating *summa cum laude* in 1923, Baxter went on to earn his Master's degree in 1925. English literature was becoming his first love, which was just as well because zoology was just not panning out. Baxter was embarrassed when he later recalled his "abysmal ignorance" in teaching pre-med students how to dissect a cat.

In 1926, Baxter wrote Colton that he had been tapped to do some part-time radio work for Philadelphia's Station WOO. He seemed more amused than intimidated by the prospect, referring to himself as a "sub-caliber pedagogue" whose topics would include "split infinitives and the simple life," followed by "mispronunciations and yodeling"

On a more serious note, Baxter admitted "I don't know how this will work out. I have never broadcasted before, and I flatter myself I have NOT a radio voice... If it takes... I am to be a feature next year." He then soberly added, "I hope it *pays* something." As a radio personality he tried his hand at everything: weather forecasts, cooking recipes, and even "advice to the lovelorn."

His broadcasting career, while relatively brief, was a kind of apprenticeship that laid the foundations of his television success twenty-five years later. Baxter learned how to relate to an audience, and to transmit information in a thoroughly entertaining manner. These skills were going to be honed in the classroom as well.

It was around this time that Baxter took two steps that would forever alter the course of his life: he decided to study abroad, and also marry Lydia Morris. The Baxter family—Frank, his new bride, and his ever-present mother—set sail aboard the S.S. Leviathan on June 11, 1927. He would spend the next two years working on his Ph.D. in English literature at Trinity College, Cambridge.

Great Britain in the late 1920s was still suffering from the post-traumatic effects of World War I. Baxter noted that the "marks of the war run deep: we are talking about the little, peaking and undersized youths one sees everywhere—the generation who spent their childhood during the war years, when meat, eggs and fresh food were doled out in quantities too inadequate for growing children."

The Baxter's first child, Lydia Morris, came along in 1929, and Francis Condie Junior (technically III) followed in 1930. Though some letters show him as a proud father—"O! She is a joy!" he writes of baby Lydia—

the novelty seems to have worn off. While never mean or cruel, Baxter could not stand what he called "childishness."

"He couldn't stand immaturity," his daughter Lydia says today. "Mother, realizing this, tried to keep us from him. We were fed in our own rooms until eight or nine." The late Frank Baxter III explained, "My father was the kind of fellow it was hard to compete with. Also, since he didn't have a childhood himself, he couldn't understand his own children." There would be times when Frank III or Lydia were three or four and would do something improper, the professor would pull a face and say, "Now that was really childish!"

Lydia became a librarian, while Frank III became a career soldier. Frank III served in both Korea and in Vietnam, where he was wounded. "About a year ago" Baxter wrote in 1967, "our son fell into one of those vile "punji" stick booby traps during a 3 a.m. patrol in Vietnam, breaking bones in his lower left leg. They flew him to a military hospital in Denver in a remarkably short time, and he is now back on duty at Fort Carlson." Baxter noted with pride that his son "teaches 'Leadership and Tactics' in the non-com school—and with some success. They gave him the Army Commendation Medal and (twice) the Master Teacher's Baton."

In the late 1920s, after seven years of research and hard labor, Baxter discovered to his horror that someone else had a Ph.D. thesis exactly like his. Casting about for a new topic, he hit upon "*Criticism and Appreciation of the Elizabethan Drama, 1642-1892*." Frank admitted this was "folly for the subject was too inclusive, and spanned too many generations, to make a tidy thesis." But he stuck with it, and was finally awarded his doctorate.

The Baxters returned home from England and headed west after a few weeks. There were still some loose ends in his graduate work to attend to, and in the meantime he landed a teaching position at the University of California, Berkeley. This was a decidedly inauspicious time for a young scholar with a growing family to be looking for work. The nation was rapidly descending into the Great Depression, with the unemployment eventually peaking at twenty-five percent of the U.S. workforce.

Desperate to make ends meet ("I have to make more" he confided to a friend), Baxter began applying to other colleges and universities. The University of Southern California took the bait, with the promise that when he officially got his Ph.D., he'd win the position of Assistant Professor of English Literature. What if Baxter had gone somewhere else, even somewhere far from the west coast? He probably would not have done

TV and the Bell films, and much of documentary film history, and the fate of PBS itself, would have been vastly different than what it is today.

And so Frank Baxter relocated to the Los Angeles area and began teaching at USC. The pay was still relatively low, and Baxter had to hustle to make a decent living. He once calculated that with "extension work, canned culture for the ladies club market, pickings, prerequisites, and what not, I hope to make something like over $3,200 next year."

For all his subsequent popularity, he was forced to teach night classes and summer school well into the 1950s. Ironically, Baxter only became prosperous when he became a national celebrity, and his various television and film work brought in a flood of extra money.

Dr. Baxter was an unusual professor, especially for the staid and Depression-wracked thirties. Above all, he wanted to give his students a much better experience in the classroom than he had been given around the turn of the century. Baxter was determined to enrich, enlighten, and instill a love of learning that would last a person's whole lifetime. The professor had suffered boredom when he was in school, without a drop of humor or visual stimulation to lessen the mind-numbing tedium. He was determined to be different.

He started to build models, reasoning that "people want to see something more than a man and a book." He made several different printing presses, miniature models of machines that would have been familiar to Benjamin Franklin, or even Guttenberg. Created with painstaking attention to detail, they were functional and could print words on pieces of parchment or paper. Some were so good they could actually print little playbills.

But the pride of his collection was his model of Shakespeare's Globe Theater as it appeared in the year 1599. Built of plywood, raffin thatch, and dowels, it stood twenty-two inches high and weighed nine pounds. Every stage needs players, so he also carved some 300 little figures to help better illustrate the range and sweep of Elizabethan drama. The Globe Theater model was his signature model, an indispensable prop in the classroom and later on TV.

Baxter in the classroom was a unique, unforgettable experience burned forever in the memories of his many students. He combined a scholar's knowledge with a child's sense of wonder, mixed with a flair for the dramatic. The doctor understood that teaching is communicating. If you don't establish a firm line of communication, the message that you are trying to deliver will become inevitably lost.

Frank Baxter teaching his Shakespeare class at USC. Note the ever-present Globe Theater model. Photo: author's collection.

Frank Baxter also knew how to grab an audience's attention and keep it. Former student Alan Simon remembers his first day in Baxter's class, which was filled to overflowing as usual. "When he entered the room, a hush filled the air with respect and expectation. He walked to the podium and stood there saying not a word. The entire class waited patiently. His pause seemed like an eternity, but was probably only a minute."

Slowly, the professor lifted his arm until his hand was covering his face. The class was riveted—what was he going to do? But then, as if to break the spell, he snapped his fingers and lowered his arm. "This," Baxter announced, "you will remember for the rest of your life."

He made a very important statement with this action," Simon says today. "If you don't have the attention of the person you are trying to communicate with, communication fails. He was a master at communicating." Baxter always kept his students on their toes. Once, the professor had the students answer roll call by voicing a figure of speech instead of the usual "here" or "present." When a student named Bill Belsey responded with "She's just an accident going somewhere to happen," Baxter called "ABSENT!"

A master storyteller, he would vividly describe the society in which a poet, playwright, or writer lived before he got to the actual literature they were supposed to study. How people lived, worked, ate, drank, and amused themselves was laid out in fascinating detail. More than one former student has remarked how Baxter put flesh on the skeleton of the past, his words bringing it to vivid life.

His asides while reading were legendary. Take for example, marzipan. There's a reference to it in *Romeo and Juliet*, Act One, Scene Five. Baxter would explain that candy was very expensive in Elizabethan days, so marzipan was a sweet substitute for most of the people. "The world," Baxter would declare, "is divided into those who can eat marzipan and those who cannot."

In another instance, when speaking of Cordelia, King Lear's youngest daughter, the professor commented "All too often Cordelia is played as sort of wan, in need of a liver extract. But she is a bright, articulate lady; what she says has a barb to it." In another class, after reading the line "By the sacred radiance of the sun" he paused and made a face, saying, "There's an awful lot of *interesting* swearing in this book!" Small jokes, perhaps, but they kept a student engaged.

Above all, he had a great sense of humor, unusual at a time when university academics were supposed to be sober and pedantic purveyors of desiccated wisdom. Once, when he was doing his television series, he got a curious postcard. "Please to where I could find the Bergerman Franklin book Mr. Baxter is reving." Writing to a librarian who forwarded the card, he couldn't resist having a little fun. "I am moved by the literate and articulate quality of the audience that is moved to such response by my efforts in educational television. I will also send our friendly inquirer important information on Georg Walsington and his wife, Marta, and about Thomas Jeferstein."

Not that all was sweetness and light. Students who came in late were subjected to a withering glance and a request asking the tardy soul to take his or her seat so the class could resume. There were few repeat offenders. Though he took roll, it was mainly to satisfy the file clerks at Admissions and Records. Baxter once said it was a matter of indifference to him if a student attended or not, because if they chose to be absent it was their loss, not his.

He was a tough grader, and his tests were, as one student put it, "devilish". The first part of a final exam would be short answer, such as: "Who said 'Sin from my lips? Oh, trespass sweetly urged! Give me my sin again'—and to whom?" But if you survived the first part, the following essay section might do you in. A typical essay question would read "Discuss (a) the weaknesses and flaws of Richard II and (b) those of his qualities that might have made him successful and happy in some other age."

But perhaps the most striking thing about Baxter in the classroom was the way he combined both scholar and dramatic actor. He didn't just read plays; he acted them out scene by scene, line by line. "Students must *experience* Shakespeare," he observed, "not just read his words." Armed with enthusiasm for his subject, as well as some real acting talent, Baxter transported his students to another world.

That a middle-aged man, bald and moonfaced, could transform himself into Hamlet, Macbeth, or a score of other characters does seem to strain one's credulity. And yet there's ample evidence he did just that. The professor didn't need heavy makeup or elaborate costuming, just his own charisma and mastery of the written word did the trick. His vivid word pictures made students feel they were on the battlefield of Agincourt, or in Verona on a moonlit night under a balcony.

Memories, however vivid, can be fragmentary, so we are lucky that a *Daily Trojan* article recorded one of Baxter's last full blown dramatic lectures before he retired in 1961. As his students watched, the classroom in Founder's Hall was suddenly transformed into a fifteenth century battlefield near Shrewsbury, England.

The professor pounded on the lectern, then turned around and faced an imaginary foe. "If I mistake not," he bellowed "Thou are Harry Monmouth. I am Harry Percy." At that moment, Harry Percy, nicknamed "Hotspur," was present in the classroom, etched in high relief by Baxter's words and the student's imagination. But the professor changed again, the metamorphosis as seamless as it was uncanny. "Why, then I see a very valiant rebel of the name. I am the Prince of Wales.."

The *Daily Trojan* article recorded that "The professor, now one, now the other pulled out a new-found sword and challenged… and the fight between Hotspur (Harry Percy) and Prince Hal (The Prince of Wales) took place as two imaginary swords clashed together and two imaginary men, more real than the classroom stage on which they were fighting, lunged at each other."

While the "fight" was in progress, Baxter stepped aside to become a third character, the fat and jovial Falstaff, shouting "To it, Hal! Nay, you can find no boy's play here!" How a six-foot bald professor could transform himself into a rotund object of mirth was anyone's guess, but in their mind's eye, the students believed it wholeheartedly.

The scene was from Shakespeare's *Henry IV*, Part I, and was a typical example of the Baxter magic. His classes were so popular they were filled almost before they were announced. Actually, in an academic career that spanned three decades, he taught a wide variety of subjects, not just Shakespeare. He taught literature from the eighteenth century, the Victorian period, and even the plays of Ancient Greece. Baxter once admitted that he didn't become the main Shakespeare professor until a colleague died in the late 1940s. "I only really taught Shakespeare the last ten years of my career," he said in 1970. "Real Shakespeare scholars wouldn't even shake hands with me!"

His reputation grew, and by the early 1940s he was the most popular professor on the entire USC campus. "If you haven't taken Dr. Baxter," the school newspaper *Daily Trojan* opined, "you haven't been to college." A student poll once voted Baxter as the man who should "teach all the classes at the university." He was universally praised, and not even his more negative aspects—like being a tough grader—could lessen his appeal. "A class with Baxter cost me a *Cum Laude*" one student admitted, "but it was worth it."

While no doubt flattered by all the attention, he was also embarrassed. "I'm just a schoolmaster," he would protest. "It's the subject they're learning, not the professor. I don't want them to be aware of me." As a reporter once remarked, keeping the students unaware of their professor was one of the few things in which he had failed.

He had at least one student who became nationally famous. In the mid-1930s Thelma Ryan was a hard-working young woman, taking campus jobs and sales positions to finance her way through college. She later became Pat Nixon, wife of President Richard Nixon and America's First Lady from 1969 to 1974. Dr. Baxter was impressed with her as a student, saying "she stood out from the empty-headed, overdressed little sorority girls like a good piece of literature on a shelf of cheap paperbacks." Trust Baxter to get in some literary allusions!

In the late 1930s, Baxter began to give public readings at Christmas time. Since he gave the readings much in the same way he did his lectures, it's not surprising that they became instantly popular. The Baxter Christmas readings became an annual USC tradition, packed as usual to overflowing. In fact, by 1949 they had attracted national attention, at least enough for *Time* magazine to send out a reporter to cover the event.

The *Time* writer noted that Baxter was a "pink faced, bouncy man" who was dressed in a twenty-year-old blue suit. The presentation was lively and dramatic, and covered a lot of material from "Dickens to

Benchley, from a medieval carol ('From Far Away We Come To You') to Ogden Nash, ('Epstein, Spare the Yule Log!') to poems written by soldiers at Tobruk."

But above all, the magazine scribe was impressed not only with Baxter's performance, but also his popularity. When the 1,500-seat Bovard Auditorium was packed, at least a hundred students huddled outside around a loudspeaker in the rain to hear him. How many professors are that popular today?

The *Time* article was Baxter's first real brush with national fame, and he admitted it left him a "little breathless." In a long letter to his old friend Dr. Colton, he described the process in humorous detail. The reporter was "talking to students, listening at keyholes to lectures, and hiding in the shrubbery." The reporter finally left after a two- or three-hour interview with Baxter himself.

But the professor admitted to some anxiety about the piece. This was a highly respected, national publication after all. Would he end up looking like a pedantic fool? "I ate nothing but barley water and a little piki bread" he continued, tongue still firmly in cheek, "and spent most of the time on my knees at the Angelus temple. When the thing appeared it was so much more friendly than I had imagined that I relaxed into a sort of protoplasmic puddle."

But he was amazed at the attention he was getting. "I have received letters from every part of the globe, except Minsk, Tomsk, and a small town in southern Uruguay. It had been fun." This was just a foretaste, a prologue, of what was to come. Three years later, thanks to TV and film, he would be well on his way to becoming one of the most famous and influential academics of his time.

2
Teaching and Television in the 1950s

THE GENERATION THAT HAD COME OF AGE during World War II had been traumatized by the Great Depression, only to be relentlessly hammered in the crucible of war. But now it seemed the worst was over, and a new era of peace and prosperity was about to dawn. Millions of servicemen returned home, eager to not merely take up their old lives but to have a fresh start and a new beginning. Many of them married and started families, the ultimate affirmation of the faith they had in America's future.

And to most Americans that faith was more than justified. The United States was booming. Stimulated in part by massive military spending—we were in a "Cold War" with the Russians—American economic productivity was on the upswing. By the 1950s, America was producing half of the world's goods, and the GNP—Gross National Product—had doubled since the late 1940s.

Most Americans had identified themselves as middle class, but the 1950s prosperity made the condition reality, not wishful thinking or "pie in the sky" daydreaming. Since the late 1940s the median family income rose from $3,083 to $5,657—even adjusted for inflation, this was an unprecedented gain. The WWII vets and their spouses wanted a home of their own, so a mass exodus began to the suburbs. There, for a little money down, or a generous "GI" loan, the young couple could buy that holy grail of the postwar America, the single family home.

Optimism reigned, and with the future looking so bright it was now time to have kids. Returning servicemen went right to "work," because the upward trend in population growth was noticeable nine months after the end of the war. In 1946, some 3.4 million babies were born, twenty percent more than in 1945. When the baby boom was in full swing, from 1954 to 1964, more than four million infants were born each year. This was to have a profound impact on American society, an impact that still resonates to this day.

The lusty infant cries were music to the ears of retailers throughout the nation. Each year's bouncing bumper crop of babies was a further spur to the burgeoning economy. And when they matured, the kids would join the ranks of consumers wanting their share of goods and services. The American standard of living, already the envy of the world, was bound to rise to ever greater heights. Swept along on a tidal wave of prosperity, optimistic Americans did their part to produce ever greater numbers of future customers.

But there were a few sour notes in the swelling chorus of baby boom praise. Educators tried to point out the nation's physical infrastructure dated to another time and could not be expected to absorb such a massive influx of new students. Schools could be built, and teachers hired and trained, but taxpayers, still frugal from their searing memories of the Depression, might not be expected to provide the required largesse.

There was also the question of time. The American public might be persuaded to pay for new schools, since it was a form of enlightened self-interest. After all, they wanted the best for their offspring. But even if the money was raised—and it was—it took time to develop plans, hire contractors, and do the million and one things to make a school a reality.

Casting about for an answer to a seemingly insoluble problem, educators turned to television. Television—educational television—would be more than a solution, it would also be salvation. It was argued that television—"the box" as it was known in some quarters—would be the panacea to cure all ills. The box would be cheaper in the long run because it avoided the boom-and-bust cycle of hiring new instructors, only to lay them off when student populations inevitably shrank.

Indeed, some "visionaries" went a step further and said television itself could be the main teacher via videotaped lessons that could be widely distributed. There would still be some kind of instructor in the classroom, but his or her role would be reduced to keeping order and giving, monitoring, and correcting tests. Even better, these video "adjuncts" required less training, and so could be paid much less than regular faculty.

The very concept of educational television was a radical one because the medium was always meant for commercial purposes. To understand this, something must be said about TV's birth and its subsequent development. Strictly speaking there is no single inventor of television, but one man comes very close. Philo T. Farnsworth was the last of his kind, a solitary genius in the mold of his idol, Thomas Edison. A Mormon farm boy, he discovered he loved to tinker and was fascinated by electronics.

At fourteen, he was plowing a potato field on his family's Idaho farm when he paused and took careful note of the fresh furrows in the earth. The field was being divided in an orderly fashion, one might say dissected. He could now see a pattern in his mind's eye. The seed of an idea was planted that was going to produce more than any physical crop. The furrows reminded him of a picture that could be broken down, transmitted, and reassembled on a screen somewhere else.

By the time he was in high school, Farnsworth imagined television could be achieved by an image dissector vacuum tube that could reproduce images electronically. This could be done by shooting a beam of electrons, line by line, against a light-sensitive screen.

In the mid-1920s he moved to San Francisco with his wife and brother in law and set-up shop on 202 Green Street, in the shadow of fabled Telegraph Hill. In fact, the new lab was a little too close to Telegraph Hill for comfort. In February, 1927, a small rock slide cascaded down on the Farnsworth lab, bringing with it fence posts, a shed, and some chicken coops. The lab only suffered a few broken windows; the fate of the chickens is unknown.

On September 7, 1927, Farnsworth and his team, which now included his wife Pem, were able to transmit an image from one room to the next. By 1929 he was transmitting images from Green Street to the Merchants Exchange Building on Battery and Washington Streets, eight blocks away. This was still too slow for the profit-minded bankers and other local money men who had advanced Farnsworth loans.

One day, James Fagan—one of the investors—came by and asked to see a demonstration of this new gadget. "When are we going to see dollars in this thing?" he asked impatiently. As if on cue, one of Farnsworth's men slid an image in front of the dissector, and in the next room the glowing screen projected a large dollar sign. Fagan got his wish: he could finally see dollars in TV!

In 1930, Vladimir Zworykin, another television pioneer, stopped by the lab to see how Farnsworth was doing. Zworykin's visit was anything but innocent, because he was under orders from his boss, David Sarnoff, to get what information he could by fair means or foul. Sarnoff, powerful head of RCA (Radio Corporation of America), saw television as a threat to his burgeoning broadcast empire. Prescient as well as powerful, Sarnoff realized that television might well have an even greater impact than radio, and was determined to have the lion's share of this new entertainment venue.

When Farnsworth refused to sell his patents, Sarnoff challenged the young inventor in court. The U.S. Patent Office eventually upheld all of Farnsworth's patents after a bitter battle that lasted a grueling five years. Emotionally drained and physically exhausted by the legal ordeal, Farnsworth drank heavily to escape his continuing woes. The inventor threw in the towel in 1939, selling Sarnoff his patents for one million dollars.

David Sarnoff never looked back. Earlier, in 1926, RCA founded the National Broadcasting Company (NBC), later called the peacock network due to its colorful logo. In 1939 RCA showcased the new medium at the 1939 New York World's Fair, the same fair that was going to stimulate the imagination of a boy named Carl Sagan.

The major networks—NBC, CBS, and ABC—were already giants in radio broadcasting, so television was a natural extension of their business activities. They were competitors but their guiding principle was the same: "free" entertainment that was not quite free, because air time would be bought by sponsors eager to pitch their products to untold millions.

At first, television was the plaything of the rich, much like the automobile was at the turn of the century, before the advent of Henry Ford. Though invented in the late 1920s, the medium didn't really take off until after World War II. There were many reasons for this: sets were expensive, broadcasts were few, and stations still rare. The Depression and global conflict also retarded television's growth.

But the industry was well established by 1950, and showed every sign of continuing growth. There were four television networks in the early fifties, NBC, CBS, ABC, and Dumont, but the first three quickly established an oligarchy. The trio had a stranglehold on American television, making sure it was commercial in all respects. Sarnoff and his fellow TV moguls might be rivals but they all held these truths to be self-evident: that the American public could be both viewers and consumers, and that the desire for entertainment was so great periodic sales pitches would be not only condoned but welcomed.

But with the baby boom in full swing, minds began to change. Maybe there was a place for educational TV after all, and not just in the classroom. Harried young housewives, busy doing laundry, washing dishes, or cleaning the house, might well find "the box" an efficient babysitter as well. Maybe the kids would even *learn* something too.

Initially the networks had very little interest in educational TV. Advertisers were much more interested in entertainment shows, because that's where the viewing audience was. No fools they! "The highest ratings," CBS director Lyman Bryson explained, "have been for a long time, and probably continue to be, the comedians who can count their listeners in the tens of millions."

Bryson might have been thinking of Milton Berle, a Jewish "Borscht Belt" comedian with roots in vaudeville. In the Fall of 1948 his *Texaco Star Theater* was a ratings juggernaut, boasting an incredible 94.7 rating. That meant that when he was on television, 94.7 percent of American viewers in the United States were tuned in to watch his antics. Engineers in New York City's water system were puzzled and dismayed how water levels dropped precipitously on Tuesdays from 8 to 9 p.m. The drops were also at regular intervals. Then, the truth dawned: the water level fluctuations took place when Berle's show was running a commercial. Hundreds of thousands of people were running to the toilet!

Things looked pretty grim for the handful of educators that saw promise in television. It looked as if commercial TV would rule America. But just when things looked their darkest, two developments strengthened the educator's hand. First, the Federal Communications Commission, which allocated channel frequencies, put a freeze on their distribution in 1948 for what was intended to be a six- to nine-month period. It lasted until 1952.

Television was growing at such a phenomenal pace, especially in the northeast corridor of the country, that the FFC's three-year-old allocation plan was already obsolete. The FFC needed time to sort things out, and make sure that VHF channels were doled out on a fair and equitable basis. The Commission also eased things by opening up the ultra-high frequency (UHF) bands to commercial stations.

The second major event in the development of educational TV was the appointment of Frieda B. Hennock as the FCC's first woman commissioner. Intelligent and hardworking, she was a tireless campaigner for public education, which she called the "electronic blackboard of the future." Hennock had been a stunning beauty in her younger years, and knew how to dress in a glamorous, eye-catching fashion.

Since Washington officialdom was overwhelmingly male, Hennock used her feminine charm, vivacity, and still considerable beauty to win friends for educational TV, then abbreviated to "ETV." She was well known for her flamboyant hats, and she didn't mind resorting to tears if the occasion demanded it. She was, in short, a consummate spokesperson for educational TV.

Hennock used the "freeze" to her advantage, calling educators to testify before the full commission on the merits of educational TV. She argued that a substantial "chunk" of channels should be reserved for ETV use, a proposal the commercial networks adamantly opposed. Scandalized to the very depths of their pocketbooks—educational TV might cut into their revenues—the networks lobbied hard against her.

In the end, Hennock managed to beat the odds and score a significant triumph against the network "Goliaths." In 1952, the FCC ended the freeze and allotted 242 channels to educational TV. The way had been cleared—but it wasn't a done deal, at least not yet. Hennock had won a victory, but would it end up more pyrrhic than promising?

Many colleges and universities considered starting educational stations on their campuses but were put off by money issues. In 1954, the costs of starting up an educational station, depending on capacity and equipment, ran from about $33,000 to $750,000, with annual operating costs between $25,000 and $500,000. These figures, high today, are astronomical for the 1950s. After all, in 1953 gas was about 22 cents a gallon, a loaf of bread was 16 cents, and the average salary about $4,000 a year.

The University of Houston was the nation's first educational (later, "public") television station. President Walter Kemmerer, who became president of the University in 1953, was the driving force in the effort. There was already a flourishing radio station on campus, KUHF-FM, but Kemmerer felt there should be more. In fact, the University of Houston's problems mirrored those of the nation at large—the same issues, only in microcosm. How could the nation's schools handle the huge influx of new students?

But for the University of Houston the problem was even more immediate. The baby boom flood hadn't arrived yet—after all, the oldest were now around six years of age. But there were many World War II veterans taking advantage of the GI Bill to go back to college and learn new skills. They were the ones who were crowding into the classrooms.

Originally, the university planned a massive and costly building program to accommodate the sudden increase of students on campus. But Kemmerer had other ideas: build an educational television station instead! Lecture halls were costly and potentially obsolete. Students could watch two lectures a week on television, and then come on campus to attend a seminar and/or be tested on the material they watched.

The University seemed to like Kemmerer's proposal, and the ambitious building plans were reduced, if not shelved completely. *Broadcasting*, a publication of the period, noted that "Houston U (sic) sees TV educational station saving $10,000,000 in building costs." Who could fail to be impressed?

Kemmerer was certain televised courses would help ease the strain. The FCC granted VHF Channel 8, and a license was issued jointly to the University and the Houston Independent School District.

The brand new station, designated KUHT, went on the air at 5 p.m. on the afternoon of May 25, 1953. With a nod to truth in advertising, the very first show was titled "It's Five." The show consisted of six lovely coeds who offered "down to earth" advice for women. Topics included make up techniques, party giving, flower arranging, and blouse making, and last, but certainly not least, preparing a child for a tonsillectomy.

As the evening wore on, viewers were treated to a couple of short films titled *Man on the Land* and *The Toronto Symphony Orchestra*, then subjected to a high school geometry review. A brief segment called "News in Focus" gave highlights of local, national, and world news, as well as sports and weather. After a presentation on administrative education, the fledgling station signed off at 7:30 p.m.

The youthful exuberance of the casts and crews (virtually all were students), coupled with the novelty of the experience, more than made up for the cheap sets, occasional "deer in the headlights" stage fright, and technical glitches. The formal dedication of KUHT was set to take place about two weeks later, on June 8, 1953.

The televised ceremony was supposed to be a showcase of the future, the inauguration of a whole new era in video education. It ended up more like a comedy of errors. Frieda Hennock was present to give a speech, but was terribly upset because there was no makeup artist on hand for her KUHT debut. Some called her the "hardboiled honey," but she was down to earth enough to kick off her shoes at dinner. Unfortunately, she couldn't find them when it was time for her to give a speech, and she came off a little "short."

But it was the technical glitches that threatened to reduce the televised ceremony into an embarrassing fiasco. Dave Garroway (a famous TV commentator at the time) had sent congratulatory greetings, but the film broke. Smoke was seen to be rising, and many feared that a cable was overheating. It proved to be only a cigarette burning a cleaning tissue in an ashtray.

But then, only a few hours before they were scheduled to go on the air, the transmitter malfunctioned. As Chief Engineer Bill Davis recalled many years later, there was a "band on the top and on the bottom of the screen and white in the middle." Davis could scarcely believe his eyes, since the transmitter had been working perfectly for the last two weeks or so.

Davis did his best, but his expertise was in radio, not television. He had worked on the campus radio station, so he knew audio broadcasting equipment like the back of his hand. But the only training he had in TV production was a crash course he had taken before assuming his new KUHT duties.

The clock was ticking—time was running out. Davis and his engineers tweaked this and tweaked that, all to no avail. The monitor still showed those terrible bands. Frustrated, the engineer walked over to the transmitter and gave it a good, swift kick. Presto! The TV monitor returned to normal! Everyone heaved a collective sigh of relief, and the show went on as planned.

Public television was literally kick started into existence. In 2003, a special exhibition was created at the University of Houston to commemorate the first ETV station. One of the displays proudly showcased a pair of cowboy boots (this is Texas, after all). These were the actual boots David wore to get the station up and running.

That same dedication day, KUHT aired Psychology 23, a twelve-week course conducted by Dr. Richard T. Evans. The telecourse was successful, and supposedly attracted an audience of 20,000. From 1953 to 1955 about eight or nine additional courses were aired for college credit. Most of these courses aired at night, to better accommodate those students who worked during the day.

By 1954, there were eight new ETV stations broadcasting across the country. Four of them were university owned, including KUON,(University of Nebraska), KTCS,(University of Washington), WHA,(University of Wisconsin) and WKAR (Michigan State University). The others, based in Pittsburg, San Francisco, St. Louis, and Cincinnati, were licensed to community groups.

In those early days, program options were few. Most programs were live, which meant they vanished into the ether a few seconds after the end credits rolled. There was also kinescope, a primitive form of videotape, in which a television monitor was filmed while the show was in progress. The resolution was usually very poor, but it was better than nothing. Gaps in the schedule would be filled in by corporate publicity reels, like *Industry On Parade*, an opus created by the National Association of Manufacturers.

Televised lectures usually consisted of nervous pedagogues, chalk in hand, scribbling words of wisdom on cheap blackboards while they sweated under the searing heat of early television lights. The overall impression was underwhelming and hardly inspirational. Not everything was bad, but there was a lackluster quality that dogged public television—though it wasn't called that yet—in its earliest years.

Educational television suffered from several drawbacks: the lack of a clear mission, lack of funding, and instructors with dull and pedantic presentation styles. What *was* the mission? Teaching a regular curriculum with material from kindergarten through graduate school? But what about general audiences in their viewing areas? What could ETV offer them?

There were those who felt that you could entertain and inform the general public. Jim Day, one of the founding fathers of San Francisco's KQED Channel 9, was one of those who felt that educational TV should entertain and inform the public. Couldn't shows be produced that would appeal to viewers of all ages? Why must public stations be forced to broadcast rote lessons, just because they bore the label "educational?"

Day soon found that not every ETV general manager agreed with him. In fact, some were adamantly opposed. John Schwarzwalder, who ran KUHT, argued public stations should stick with instructional programming no matter what. That, and nothing else, was their primary mission. Schwarzwalder seemed to relish his role as gadfly and all-around naysayer, dubbing himself, at one point, the "vice president of dissent." Arguments grew heated at times, with Schwarzwalder sarcastically suggesting that Day and Ford Foundation rep Fred Friendly "kill themselves to rid educational television of its general audience advocates."

Day and his colleague Jonathan Rice faced challenges even more daunting, in large part because of the peculiarities of California state law. They could not get money from schools or school districts because it was illegal. Deprived of a natural source of funding, KQED was forced to be a "community" station from the very beginning. When "seed" money from the Ford Foundation and other similar sources ran out, KQED became desperate. Eventually the station raised money by televised auctions—a technique so successful it's used regularly to this day.

ETV was also a victim of its own sales pitch. Television was supposed to be cost effective, a cheap, viable alternative to real teachers and real classrooms. The early EDT stations were pioneers and knew it, so they worked extremely hard to present a viable product. But shoestring budgets often meant shoestring organizations with slapdash facilities, quickly outmoded equipment, and revolving-door staff who left seeking better opportunities and better pay.

KUHT, for example, never lived up to its bright promise, though there are some who would dispute this assessment. The station was located in the fourth largest city in the United States and had access to all the oil tycoons in the region. It should have become a powerhouse like WGBH in Boston but it never quite made it. For the first ten years of its existence, its transmitter was underpowered and, as the years went by, became outdated.

The Houston station kept a low profile for many years, quietly stagnating while other stations moved on to glory. This gave rise to what some KUHT staffers called the "Channel 8 disease." Though rarely fatal—except to the station's reputation—its symptoms included frustration, boredom, and resumé writing in hope of abandoning ship.

Because ETV was non-commercial, it was heavily dependent on public-spirited millionaires or institutions like the Ford Foundation. Sometimes this generosity came with strings attached. The most notorious example of this was KTHE-TV, Channel 28, an educational station founded a few months after KUHT in Los Angeles. Captain Allan Hancock, a member of the University of Southern California Board of Trustees, bankrolled the project, which would be located on the USC campus.

Once the station was up and running, viewers were treated to an act called the Hancock String Quartet, with our old friend Captain Hancock playing the lead violin. Tired of playing second fiddle to his String Quartet, the Board of Trustees disagreed with him on the direction the station was going. When the arguments grew too heated, Hancock withdrew his financial support and KTHE folded after only a few months. The irony, of course, is that USC was where Dr. Baxter was teaching at the time.

Sometimes a "classroom" set would be built in the studio, and sometimes a camera would simply record a teacher's presentation or lecture in a real setting. It didn't really matter, because too often the presentation was boring, a Sahara of dry, mind-numbing prose delivered in a monotone. The instructor's body language said it all: he or she was usually stiff, ill at ease, and unenthusiastic about the subject.

Educator Roger P. Smith once recalled witnessing such a TV show with a classroom of students. "The teacher," Smith recalled, referring to the video image, "stood at the chalkboard and gave the lessons as usual.

The dullness that children were required to watch raged from painful to excruciating." Smith also confessed that even as a professional "I managed to stay awake for only about five minutes."

A report issued by the Educational Television and Radio Center tried to be more diplomatic, saying "We (ETV) have preferred not to sacrifice content values for performance, and at times this has resulted in programs that lack the vitality and excitement desired." This was the understatement of the decade!

Though there were exceptions, most ETV officials stuck doggedly to their pedantic formats that produced a bumper crop of lackluster programming. The basic truth that learning can be fun, and even attract audiences if presented the right way, was lost to many in public TV.

One of KUHT's earliest programs showed the way—but its example was largely ignored, at least at the time. Once a month the station would live broadcast a meeting of the Houston Independent School District Board. The very concept sounds dull, but the presence of a camera seemed to encourage grandstanding and partisan debate. The arguments grew louder, the sessions longer, and spectators in the board chamber added to the pandemonium by cheering and booing to their heart's content.

Before long, the School District Board meetings were attracting 100,000 viewers, who described the goings on as "the best programs on the air" and "more comedy than Milton Berle." The viewing audience also learned something as well, though this was not asked when they were polled.

They were entertained, but while they were laughing, people learned about the functions of a school district, where their tax money was going, and other important facts. The humor helped people swallow dull and unpalatable facts. You *could* be educated and entertained at the same time, but finding the proper mix of these elements was the tricky part.

While educational TV limped along, a medium unsure of its direction and starved for funds, commercial TV was breaking new ground in cultural, educational, and public affairs programming. One of the prime movers in this development was the Ford Foundation, which had lent its support to KUHT and other fledgling public stations. But the Foundation also strongly believed that commercial television had a role in education as well.

To this end it retained the services of James Webb Young, and asked him to do what he could in this direction. Young was not sure that this could be done—that commercial networks could be persuaded to broadcast culture and learning as well as mindless entertainment—but was persuaded by talks with such industry leaders as NBC mogul Sylvester "Pat"

Weaver. Weaver even claimed that it wasn't even necessary to set aside so many channels for noncommercial educational television, because the commercial networks would do their part in bringing quality learning to the viewers,

One of the first things that Young did was to hire Robert Saudek to produce a show that came to be called *Omnibus*. Young chose well, because Saudek was a man of integrity, skill, and artistic sensibility, with a genuine commitment to excellence.

When *Omnibus* premiered on November 9, 1952, some three million viewers tuned in to see a forty-four-year-old Englishman named Alistair Cooke welcome them to a new series that held "something for everyone." Cooke, better known for his later *America* series and his stint on *Masterpiece Theater*, was the host for *Omnibus*'s eight year run.

The shows indeed had "something for everyone," with excerpts from classical and contemporary drama, music, and dance. There might be a scene from Gilbert and Sullivan's *The Mikado*, an original play written for television by William Saroyan, or a selection from Shakespeare. Speaking of Shakespeare, Saudek scored a coup when he persuaded Orson Welles to play the lead in an abridged, live version of *King Lear*. First aired in 1953, this *Omnibus* episode can be purchased on DVD.

In between the larger segments were the small features, a buffet of subjects that held the viewer's interest. There might be a fascinating segment on X-rays, how a jackrabbit runs, or a piece featuring oceanographer Jacques Cousteau. If one segment wasn't a viewer's cup of tea, all he or she had to do was to stick around—there was always something of interest somewhere on the show.

Omnibus showcased talent, and, for many, an appearance on *Omnibus* set them on the path to stardom. This list includes such people as Joanne Woodward, Jonathan Winters, Ed Asner, Larry ("J.R.") Hagman, and James Dean. Established stars also did segments, including Hume Cronin, Jessica Tandy, Bert Lahr, and Charlton Heston. Since the performances were live, the actors were given a "stage play" atmosphere, but with audiences of millions.

Omnibus had a big budget, but it was Saudek's genius that help *Omnibus* stand apart from all the rest. For example, in the episode "Dancing Is a Man's Game," pro-golfer Sam Snead demonstrated the perfect golf swing. Next, a Wimbledon champion showed the perfect tennis backswing, followed by a skier balancing. All these elements were combined in a dance choreographed and performed by movie legend Gene Kelly.

There were many non-actor guest "stars" on *Omnibus*, including the poet Robert Frost and a somewhat curmudgeonly architect Frank Lloyd Wright. Dr. Frank Baxter made an appearance on the November 29, 1953 episode, armed with his ubiquitous Shakespeare's Globe model. Baxter and Alistair Cooke were perfectly matched, both being witty and erudite men, and the segment is noted for a little good-natured verbal sparring. But Baxter also gives a very entertaining mini-lecture on the Globe, a visual record, jokes and all, of what he must have presented to his students.

Omnibus was broadcast during television's Golden Age, but, like all golden ages, it did not last. The Ford Foundation withdrew its support after five years, though Saudek managed to keep it afloat until 1961. The main problem was the ever-changing face of commercial television. Profit and the "bottom line" became more important than education and culture, and the "mindless" entertainment shows attracted both sponsors and viewers.

But there was still hope for entertaining education. The concept was still a possibility, but needed the proper "midwives" for it to be born. The same year that KUHT was founded, Dr. Frank Baxter made his television debut and took the country by storm. Both public and commercial educational TV would never be the same.

But his influence was going to have an even greater impact when he starred in a series of science films produced by the Bell System. These Bell Science films were a template, a foundation for documentaries that were to come. In fact, the sheer impact the Bell films had on educational series and science documentaries was profound, and resonates to this day.

3
All the World's a Stage, 1953

The year 1953 was crowded with newsworthy events: Dwight D. Eisenhower became President, Communist dictator Joseph Stalin died, and the Korean War finally ended after a bloody stalemate of three years. Television was gaining popularity because the programs were free and you could watch them from the comfort of your own home. Post-war prosperity meant more money for consumer goods, and televisions were high on the priority list for most Americans.

That December, there were 360 television stations in the United States, with 231 of them having just started within the year. In this period, TV Guide began publication, the first TV dinner made its appearance, and the first commercial color program, the *Colegate Comedy Hour* with Donald O'Connor, made its debut.

As the decade progressed, Americans were buying television sets at the rate of 20,000 a day. They were becoming the new status symbol. As soon as a family brought a new TV, they found themselves besieged by neighbors eager to catch a glimpse of the new medium. As the word spread, even complete strangers as far as two or three blocks down the street might show up at your door.

I Love Lucy, the now-classic sitcom featuring Lucille Ball and Desi Arnaz, still ruled the airwaves. On January 19, 1953, an estimated sixty-eight percent of all U.S. television sets were tuned in to watch Lucy give birth to "Little Ricky." But there was criticism too. Long before FCC Chairman Newton Minnow made his famous "TV is a vast wasteland" speech, modern day Jeremiahs were warning against the new medium's mindlessness and banality. The violence of cops shows and westerns would create a generation of amoral juvenile delinquents. And even if they did escape the "Graybar Hotel," they risked becoming overweight couch potatoes by sheer "glued to the tube" inactivity.

Of course, compared to the garbage we see today with four letter expletives, toilet humor, and graphic sex and violence, even the worst '50s shows seem classic. In part because of viewer demographics—modern networks pander to the more-or-less college-age crowd—many shows have their mind in the gutter. Today, even the title of the classic show "Leave it to Beaver" would be the subject of double entendre jokes.

Columbia Broadcasting System (CBS) Los Angeles affiliate KNXT offered the University of Southern California an hour of "public service time." Sounds very generous, except the period selected was 11 a.m. on Saturdays. The time slot was obviously a backwater, a place where the station could appear high-minded without sacrificing the advertising revenue of later, more commercial periods.

The university asked Dr. Baxter if he'd like to accept the assignment. He wasn't sure. Saturday morning was the exclusive domain of children's programming, the land of such kiddie fare as *Howdy Doody*, cartoons, and the *Roy Rogers Show. Romeo and Juliet* might be competing with *Rootie Kazootie.* (A character who played a kazoo—let's not go there!) Baxter felt the only audience he'd get was "two librarians and a bedridden man."

The overall effort was called Southern California Summer Session, even though it ended up being telecast in the early fall. Dr. Baxter's show was titled *Shakespeare on TV*, a name that invokes both the content of the programs and their modest aims. Dr. Baxter would present eighteen lectures on the Bard, each about forty-five minutes long. The course could be taken for one unit of credit, audited, or simply viewed for one's one personal enrichment. The first series covered such plays as *Romeo and Juliet, King Lear, Henry V, Othello*, and, everyone's standard, *Hamlet.*

His KNXT-TV "classroom" was pretty basic—a chalkboard, a camera, and about fifteen chairs for a small live audience. In fact, it was the same low budget, primitive scenario that other professors had endured before him without recognition and certainly without success. The idea, at least in theory, was that Baxter was going to replicate the teaching he did on the USC campus. Of course, the Spartan set was easy on KNXT's budget as well—always a primary consideration.

Frank Baxter's first show aired live on Saturday, September 26, 1953. He spoke from a lectern, with occasional asides to a tripod that held some pictures or to a blackboard that had the logo KNXT-TV running along its top. There were no rehearsals. The TV director, one William Acine, would call for camera changes when he sensed the need—like when Baxter suddenly changed pace, hesitated for a word, or looked down to read another portion of Shakespeare's text.

Some 400,000 tuned in to watch the programs, 350 paid $12 to take the "class" for credit, and 900 audited. A second series was telecast, and a third. He soon was attracting audiences of 750,000, and his ratings were higher than some commercially produced shows. His new video students ranged from sixteen to ninety-one years of age.

The series was such a success it was continued. In a typical season, for example, *Shakespeare on TV* began on Saturday Oct 1, 1955, and ended on January 21, 1956. For those who took it for credit, a two-hour final was given on January 28, from two to four o'clock, in Founders Hall on the USC campus.

A 1955 semester brochure gives us an idea of what the course "syllabus" was like. The 1955-56 edition of *Shakespeare on TV* included a presentation on "The World of William Shakespeare," followed by lectures and readings on *Richard III*, *Macbeth*, *Antony and Cleopatra*, and *The Tempest*. Standard material, but it was Baxter's entertaining side comments, dramatic interpretations, and ready humor that made his offerings so popular with students and viewers alike.

For example, in one lecture, Baxter urged his audience to envision the famous balcony scene in *Romeo and Juliet* as it was performed at Globe Theater in 1599. "The actor had to stand surrounded by his audience," Baxter began. "There were (even) people seated *on* the stage. He (Romeo) had to say 'Excuse me, my lord, excuse me Ed, Hi Bill—Oh, Juliet...'"

When pressed, Baxter admitted teaching on TV was an art. "My business is informative entertainment," he once declared. "I'm in an entertainment medium, but I can still hold attention by the wonders of information." In another interview, Baxter explained "The TV teacher needs stage presence, ability to communicate... and a very strong sense of the visual."

He did indeed have a strong sense of the visual. He brought along his usual props, the ones he had been using for the past twenty-odd years: the Globe Theater model, a drawing of Elizabethan London, a map or two, and plaster bust of Shakespeare. He knew that, cliché or not, a picture really is worth a "thousand words." Baxter moved on to other shows, other concepts, but he continued his *Shakespeare on TV* course for a few more years.

The good doctor had an immediate effect on the public. His campus office, which he jokingly referred to as "Cell Block X," was flooded with hundreds of letters, a pile of correspondence that grew along with his fame. In a typical missive, a mother wrote Baxter that when his program came on the air her fourteen-year-old son and his friends came in off the street to watch the show. "You have interested him in reading for the first time in his life," the awed mother explained.

Another letter came from a small business owner, who said she shut up shop on Saturday mornings and invited her friends to see Baxter's show over cake and coffee. His appeal was universal. People of all ages and walks of life tuned in regularly, and then spread the word to their neighbors.

Many letters were from women—usually "housewives" who had left high school or college to get married and have children. This too was a "characteristic" of the Eisenhower '50s; a woman was supposed to abandon any career ambitions in favor of being a wife and mother. It was a man's world "out there," and society said she had to confine herself to domestic duties on the "home front." The housewives who wrote Baxter were grateful that a bit of the longed-for college life was coming into their living rooms, even if it happened while they were ironing or doing the laundry.

Dr. Baxter was just as surprised as anyone else by the overwhelmingly positive response. His office inbox at the University became choked with mail, and more letters came with every delivery. The KNXT switchboard was also deluged with phone calls. "It's mind-boggling," Baxter said in awe, "to know that I've been a teacher all my life, yet now on a given Saturday morning I teach more students than I have than in all my 23 years in the profession."

By December 1953, only four months from his debut, Baxter was already gaining national attention. *Life* magazine, then the most prestigious periodical in the country, did a profile in their December 7, 1953, issue. "TV and Teachers Team Up," proclaimed the banner headline, and the article praised Baxter as being "erudite and witty." The magazine was awestruck by his performance and the sheer scope of his influence, noting, "When Baxter raps for attention, some 750,000 persons come to order."

Life also hinted that shows like Baxter's might be salvation for the nation's new and struggling public education stations. The magazine noted that the country's second all-educational station, KTHE, was about to start broadcasting that same month, and Baxter's Shakespeare shows are already a feature on the schedule.

Dr. Frank Baxter teaching his *Shakespeare on TV* course. Photo Courtesy University of California, USC Libraries Special Collections.

The "Baxter phenomenon" spread across the country, radiating out from Los Angeles like ripples travelling across a placid pond. Programs like Baxter's were live but were also recorded in a primitive form of recording, called kinescope. It was a simple but effective technique for the time. Basically, as mentioned in a previous chapter, a camera filmed a TV monitor as the live show progressed.

These kinescopes were often grainy and a little blurry, and of course they were usually shot in abysmal, poorly lit black-and-white. Yet, even against these odds, Baxter won the day by charisma and sheer force of personality. It seemed people were hungry for his kind of program.

His first big "conquest" outside was the all-important New York market. It's no exaggeration to say he took the Big Apple by storm. Among the first to be won over were the New York critics—a hard-bitten lot known to almost take pride in their cynicism. *New York Times* critic Jack Gould was one of the first to sing Baxter's praises when the first kinescopes arrived in New York. The first one aired June 12, 1954, with a discussion of the development of the Elizabethan theater, from the old English inns to the Globe Theater.

Armed with a long pointer and his ubiquitous Globe model, Baxter literally and figuratively set the stage for the rest of his series. "With humor and detail he described the limitations and virtues of the stage, and their influence on the audience, on the actors, and on Shakespeare," wrote Gould. "It was all as if the setting suddenly had become real, human, and meaningful." Harriet van Horne of the New York World Telegram and Sun concurred, saying "This is not ersatz, sugar coated education. This is real!"

The New York market aired *Shakespeare on TV* on Saturdays from 2:45 to 3:30. This was hardly an improvement from the LA experience. Late Saturday afternoon wasn't exactly prime time, and this was compounded by the fact many New Yorkers were on vacation.

Gould was dumbfounded, writing, "Here is probably one of educational TV's first smash successes; it has won more than ten national and regional awards. And yet, what has happened to it?" Expanding on this theme, he said "The simple economics of commercial television are keeping Dr. Baxter out of a more civilized hour when his audience could be manifold what it is now. There is choice time for sponsored junk, but not for a sustaining jewel."

The scheduling debate aside, Gould came close to explaining the Baxter magic when he said the professor's approach "is not that of an assignment in literature. Rather it is an adventure to which he brings a sense of enthusiastic discovery that he delights in sharing with his student or viewer."

Baxter's first appearance was on commercial TV, but his Shakespeare shows quickly became a godsend to many of the new public stations, giving them a transfusion of excitement that was lacking in their other programming. San Francisco's KQED is a prime example of how Baxter and his programs could resuscitate a station.

In some respects, KQED's birth was even more precarious than that of KUHT. A handful of librarians, school superintendents, and museum directors formed a committee somewhat grandly known as the Bay Area Educational Association to try and get an educational station on the air. As a first step, they hired Jim Day and Jonathan Rice to try and get the ball rolling.

Funds were so scant that KQED's first office was in the back of Day's station wagon. It was Day's wife Beverly who came up with the name KQED. It's an acronym for the Latin "quod erat demonstrandum," or "which was to be demonstrated." But KQED needed more than just a catchy name if it was going to start broadcasting. The Ford Foundation chipped in with some money, and donations from all sources—even bake sales—raised enough to begin operations.

KQED managed to get commercial station KPIX-Channel 5's old studio and transmitter in the Mark Hopkins Hotel atop famous Nob Hill. The financial situation was shaky, especially since it was learned that California state law prohibited schools from using funds for television education. Nevertheless, KQED officially began with a test pattern that was telecast in April of 1954. The broadcast schedule was a grand total of two nights a week, airing such titillating fare as "*Our Friend the Atom*."

But Frank Baxter's *Shakespeare on TV* was just the right medicine to help an ailing station that was fighting to survive. KQED began airing the Shakespeare series in the late summer of 1954, and it predictably became, as the station put it, "a smashing success." *Focus*, the in-house KQED magazine, noted the "Enrollment at Mill College for his *Shakespeare on TV* series reached a total of 431, with seventy-seven taking the course for one unit of credit. Many women's organizations have joined registrants scattered over thirteen counties of Northern California in forming 'coffee classes' and regular study groups to view the series."

In the meantime, KQED was experiencing some growing pains and needed new facilities. The San Francisco School District leased the John O'Connell Trade School and started converting it to a studio. A control booth was added, and everything was made as soundproof as possible. Signs were posted in the restrooms just above the studio with the warning "Don't flush during broadcasts."

The new studio was inaugurated on November 8, 1954, a date which coincided with the launch of a new Baxter Shakespeare series. Baxter came up from Los Angeles to do a live *Shakespeare on TV* episode as part of the overall dedication ceremonies. It was a wise choice to include Baxter, whose mere appearance guaranteed an audience. Stills taken at the time show Baxter pounding on a lectern to make a point, while a large studio audience looks on enraptured. A chalk board and a bust of Shakespeare are the only other props.

The first live program ever broadcast by KQED was Jim Day's *Kaleidoscope.* It was almost a forgone conclusion that Dr. Frank Baxter would be the show's first distinguished guest. Baxter was a natural—because his shows were so popular people were going to tune in to see and hear him. Dr. Baxter helped launch *Kaleidoscope,* which became a KQED mainstay for many years. Other guests included such luminaries as Eleanor Roosevelt, Bing Crosby, Robert Kennedy, and Aldous Huxley, the same author who would later submit a script for Baxter's *Our Mr. Sun.* But the growing popularity of Baxter's *Shakespeare on TV* caused some to wonder why such quality programming—even if on fuzzy kinescopes—should be allowed to languish in odd times.

In the 1950s, academics were supposed to be remote, almost "otherworldly" intellectuals whose thought patterns were incomprehensible to the average mortal. Occasionally they might descend from "Mount Olympus" to lecture the common herd, but they were supposed to be achingly dull and totally boring. Given these attitudes, Baxter was a revelation. He may have *looked* like an academic, with his wire-framed glasses, bald pate, and grey-flannel suit, but he didn't *act* like one

Baxter's great contribution to TV education was the way he made academic subjects interesting and accessible to the general public. He had the attributes that would later serve presenters like James Burke so well. Dr. Baxter had a thorough knowledge of his subject, but he also had a genuine enthusiasm and love for what he was trying to teach. This enthusiasm was contagious, as anyone who has seen a Baxter program can readily attest. It "infected" the viewer with an intellectual curiosity and thirst for knowledge that would last a lifetime.

He knew the value of self-deprecating humor and used comic asides to deflate the public's perception that Shakespeare was too "highbrow" for the average person—at least the average American. He was the consummate presenter and ideal for the new visual medium. Frank Baxter combined a scholar's knowledge with a well-honed actor's flair for the dramatic.

A mere six months after his television debut, Baxter found himself a contender for an Emmy. On February 11, 1954, a bespectacled Baxter, dressed in a conservative suit, found himself sitting among Hollywood royalty. The sixth annual Academy of Television Arts and Sciences award show was held at the Hollywood Palladium on Sunset Boulevard. Over 1,300 people were in attendance, including Lucille Ball, Desi Arnaz, Donald O'Connor, and a host of other performers. Ed Sullivan, the famous "really big sheew" variety host, was on hand as the Master of Ceremonies.

To Baxter's astonishment he won not one but two Emmys that evening—one for the best public affairs program, and the other for outstanding male performer. Coming up to the podium to accept his statuettes, Baxter looked at the golden prizes with a mixture of disbelief and wry amusement. "I'm sorry to announce," he began, tongue firmly in cheek, "that the person who deserves this award—William Shakespeare—cannot be here, due to a long absence. I would, however, like to thank my writer who, by strange coincidence, is also named William Shakespeare." The audience roared; the moon-face professor had brought down the house.

In the wake of the overwhelming success of *Shakespeare on TV* KNXT and USC quickly made plans for another fifteen episode edition. Baxter was something of a hot property, and KNXT wanted to strike while the ratings "iron" was hot. The result was *Now and Then*, which began on August 1, 1954, and ended in May of 1955. Shakespeare was still read and discussed, but Baxter broadened his scope by adding Anglo-Saxon literature, Walt Whitman, John Donne, and other writers and poets of interest.

Baxter's presentation style never varied: He'd read from selected works but pause to explain a passage, provide background, or even make a wry comment or joke. The Globe Theater model made its customary appearance, but was supplemented by a Viking ship and other props.

Another popular favorite was his show *The Written Word*, which was produced for the Educational TV and Radio Center in Ann Arbor, Michigan. This was an "in house" production, lensed on campus at USC's cinema department. He began with the development of the alphabet and written symbols, starting with Egypt of the Pharaohs. The Phoenician, Greek and Roman contributions are examined.

The topics sound sleep inducing, but once again the Baxter magic makes them spring to life. "As you see," he says at one point, examining an inscription from 1300 B.C., "A, K, N, O, and T haven't changed much in the past 3,000 years. But look at the M and N—always the little squiggles!"

Aside from charisma, it's his visual hands-on approach that sets him apart from his video contemporaries. When he speaks of Egyptian hieroglyphics, he's not just content to talk about papyrus—he makes it on

Dr. Baxter explains ancient Egyptian hieroglyphics in the series *The Written Word*
Photo courtesy of the University of Southern California, on behalf of the
USC Libraries Special Collections.

camera. Taking some southern California fibrous substitute—the original is extinct along the Nile—Baxter cuts strips and then pounds the strips with obvious relish.

Baxter takes the flattened fibers, cross thatches them, and puts them aside to dry. Smiling from ear to ear, the doctor promises in a later episode "we'll make paper." It was this episode, first telecast in the summer of 1956, that got me "hooked" on Dr. Baxter.

More accolades and honors followed, including the Sylvania Award in 1953, a Peabody Award in 1956, and the Toastmasters International first Golden Gavel. His programs also won five more Emmys, bringing the grand total to an amazing seven in all. Brandeis University honored him for "reawakening interest in the classroom," and USC and three other colleges gave him honorary degrees.

The professor was an intensely private man, who preferred such solitary pursuits as reading and model building. He was not a party goer or "back slapper," though he did attend a Hollywood event or two, and he rubbed shoulders with the stars with obvious pleasure. Of course, fame was a two-edged sword. It's one thing to be a popular professor on campus—this he could understand—but it's quite another thing to be a national celebrity with a growing roster of prestigious awards and fan mail that a movie star would envy.

"There are some stores I can no longer patronize," Baxter once wrote, "for I seem to have...a reception when I stand still." He did enjoy himself, and the extra money meant he didn't have to teach night school or sum-

Dr. Baxter with one of his famous models, a working printing press
Photo courtesy Author's collection.

mer school to make ends meet. He was also able to build a swimming pool at his home at 1614 Camden Parkway, South Pasadena, California.

And then, a curious thing happened: he started going into acting and hosting commercial television shows. This is something that later famous academics, like Carl Sagan, never tried. Immediately after his first two Emmy wins, he was contacted by Lucille Ball, who was then at the height of her television comedic fame. She had been in the audience when Baxter received his awards, and she was highly impressed. Ball invited Baxter to do a guest shot, but he declined with a bad pun, explaining "I love lucidity!"

This sounds snobbish, but he was very picky about the projects he accepted. One of his first acting "gigs" was an episode of *The George Burns and Gracie Allen Show*, which aired in 1957. Dr. Baxter played himself, which was the ultimate acknowledgement of his star status. In a kind of art-imitating-life scenario, the Burns' son Ronnie is taking Baxter's Shakespeare class at USC. One of Ronnie's toughest assignments is to do an essay on the Bard.

Scatterbrained Gracie wants to help, so without her son's knowledge she rewrites the essay, with predictable comic results. The episode was a success, and led to other acting jobs. He also became popular as a host/narrator in a number of shows, most notably the *Bell Telephone Hour* and *Four Winds to Adventure*.

Baxter also did a very funny series of skits on the *Tennessee Ernie Ford Show*. At first, this doesn't seem logical. Tennessee Ernie was a good singer with a substantial following, but *I Love Lucy* was the highest rated show in the land. It seems that Tennessee Ernie's folksy sayings were very appealing to the professor. "Ernie's use of language is almost Shakespearian," Baxter declared. "If Shakespeare were alive today he would be proud to have written such a line as 'He's as nervous as a long tailed cat in a room full of rocking chairs.' This is truly brilliant in the imagery it evokes."

Tennessee Ernie Ford's program was an eclectic mix of comedy, dance routines, and, of course, singing. Ernie possessed a rich, deep voice, which he used to great effect. His most famous song was "Sixteen Tons," an unvarnished paean to the woes of the working man. Ford, who was a very religious man, usually ended each show with a spiritual.

Dr. Baxter was the guest star on the June 20, 1957, program. Baxter comes on stage with a set of bongo drums under his arm. "It's a great honor for us to have you here!" Tennessee Ernie declares, to which the professor replies. "Thanks a mil, Daddy-o!" Taken aback by the hep talk, Ford asks why he's adopted this way of speech.

Baxter and Bongos—Baxter with singer Tennessee Ernie Ford, *The Tennessee Ernie Ford Show* June 20, 1957. Photo Courtesy Tennessee Ernie Ford Enterprises.

"Well," Baxter explains "the teaching of the Bard is a luxury I can't afford." He notes that entertainers are paid a lot more than college professors. He's not earning a lot of money, or as Baxter puts it, "I'm not making it, man!" As the routine continues, Baxter explains he's going to sing folk tunes, accompanied by the bongo drums. He's all set, and he even has an agent that has him booked at a county fair once the semester is over.

Ford warns Baxter that the show-business life is not what he thinks it is. This sets up a series of hilarious skits: In the country fair segment, Baxter is constantly interrupted by Ford, posing as a woman with a pie, a barker, and a hayseed with a live baby piglet. The next segment shows Ford as a disc jockey named Rex Oedipus. Once again, Baxter is comically interrupted by a wise-cracking Ford in a tacky "radio studio" set.

The funniest bit was when "Rex Oedipus" produces a huge lollipop to do an on air commercial. The product is an "all day sucker," and as Ford talks, Baxter begins to lick the lollipop. He stops when Ford glibly announces that the flavors include strawberry, lemon, and sheep dip, and "yours is the *sheep dip!*" The look on Baxter's face is priceless.

In 1956, Dr. Baxter appeared in *The Mole People*, a science fiction movie produced by Universal Studios. It starred John Agar as archaeologist "Dr. Roger Bentley," who, with his partner Hugh Beaumont ("Jud Bellamin"), stumbles quite by accident on an underground race of ancient Sumerians. They've been troglodytes for so many generations these sub-

Baxter samples the "sheep dip" flavor lollipop while Tennessee Ernie looks on.
From the June 20, 1957, episode of *The Tennessee Ernie Ford Show*.
Photo courtesy Tennessee Ernie Ford Enterprises.

terranean Sumerians are all albinos. This hidden civilization is supported by mole people—mutant humanoids who look something like the Sand People of the *Star Wars* pictures.

Agar and Beaumont are initially welcomed as gods, mainly because they possess the "fire of Ishtar"—namely a large flashlight. Agar is given a slave girl who is a throwback—she has normal skin pigmentation. Before you can say "ziggurat," Agar and the girl fall in love. But the High Priest Elizar, played by a very British Alan Napier, plots against the archaeologists.

The mole people, who have been brutalized and oppressed for centuries, conveniently revolt, allowing for our heroes to escape to the surface. Originally the slave girl was to have survived, happily going off into the sunrise—her very first!—with her man. But this was the 1950s, the Civil Rights movement was brewing, and Universal didn't want to offend the southern (read: white) market.

So the actors were called back after two weeks and the final scenes were reshot. There's an earthquake, and she inexplicably runs back into the cave entrance and is crushed to death. What was all the fuss? She was an albino woman who could "pass" for surface human and was about to marry a surface man. This smacked of miscegenation, of "race mixing," and so had to be changed

Mole People is now a camp classic, an entry in the "so bad it's good" hall of movie infamy. The sets are cheesy, the art direction poor (there are ancient Egyptian hieroglyphs in the Sumerian temple), and the dialogue banal. John Agar was a decent actor who was famous mainly for two things: for being in John Ford's classic cavalry picture *Fort Apache*, and also for being Shirley Temple's first husband. *Mole People* was yet another nail in his career's coffin.

Some of the other cast went on to better if not bigger things. Hugh Beaumont achieved a kind of video immortality as the father in *Leave It to Beaver*. Alan Napier played many roles, but he was most famous as the butler in the *Batman* series starring Adam West.

There are some mysteries that seem to be unsolvable, like the meaning of life, the origins of the universe, and why Dr. Baxter chose to appear in this cheesy production. It couldn't have been *only* for the money. He shows up even before the credits roll, in a studio mock-up that's supposed to be his office at USC. Baxter gets up from behind a desk and begins a mini-lecture on various hollow earth theories through the ages, which runs about four and a half minutes.

Baxter, playing himself, gives a very serious, almost "deadpan" lecture. He walks over to a large mounted globe, taps it with his finger, and asks "What's inside this globe? What is beneath our feet?" The professor goes on to say that most of the earth's surface is mapped and known, and we are just starting to probe into the depths of space. (This was a year before Sputnik and the beginning of the space race.)

After a brief review of ancient beliefs, he talks of some nineteenth century crackpots. There was John Cleves Symmes , "a soldier, and rather minor hero of the war of 1812," Baxter explains. "He had this sudden idea that beneath our world were globes within globes, five of them, and some of them inhabited. If you went to the frozen wastes of our world, like in Siberia, you could find a hole that would lead you down into them."

Unfortunately for Symmes he was so obsessed with his theory he scrambled around lecturing on the subject, only to succumb to "fatigue, and he died before he could make his experiment." By this time the movie audience is starting to envy Symmes, because at least he was out of his misery. They still had another two minutes of lecture to endure. Baxter continues the tale, with more theories and more "giants" of hollow earth "science."

Today, film critics pillory Baxter as the "gesture professor," due to his constant arm waving, lapel grabbing, and so on. These were some of his real mannerisms, but these critics miss the point; Baxter's mock solemnity and arm waving are trying to show that this is an in-joke, a joke he is trying to share with the audience.

As the segment wraps up, Dr. Baxter says "The movie is a fable, beyond fiction, for I think that if you study this picture, and think about it when it's over, you'll realize it's more than just a story told, it's a fable with a meaning and a significance for you and for me in the twentieth century." Granted, this movie is not one of Baxter's finest hours, but research shows much of what he said was based on actual fact.

Baxter also played a role in *Mr. Novak*, a popular drama about a high school English teacher. James Franciscus played the central role as John Novak, and character actor Dean Jagger was Principal Albert Lane. Baxter was a guest star in the episode "X is the Unknown Factor," which aired in October of 1963. The professor played "Dr. Gagan," who comes to Novak's High school to interview students for an important scholarship. Actor David Mackin played "Mike Daniels," a student who's up for that scholarship.

As the drama unfolds, Mike realizes he must pass Mr. Novak's English final or he will not even qualify for the coveted prize. He cheats, and he is later interviewed by Baxter in the guise of "Dr. Gagan." Mike wins the scholarship, but he has second thoughts and a guilty conscience. The lad confesses his cheating at the end of the show. He loses his chance at a scholarship, but learns a lesson and regains his moral compass and self-respect.

It's often said that everyone has a twin somewhere in this world, and Frank Baxter found his on the set of *Mr Novak*. Character actor Dean Jagger was a near mirror image of the professor, so much so that publicity photos underscored the uncanny resemblance.

Only seventeen at the time, Macklin has warm and indelible memories of Frank Baxter. Like most "boomers," he saw the Bell Science shows and had become a fan. "I was excited to meet and work with him" Macklin says today. "I found him affable and interested and he smiled and had the twinkle in his eye. I told him how much I enjoyed his work, and he said 'Thank you, David, I enjoyed yours.'"

This was no courteous chit chat—Baxter really did know of the young actor's previous roles. "I was pleased to learn, "Macklin continues, "that he had seen the award-winning PBS documentary '*Escape from a Cage*,' about youthful mental problems. He also had seen the previous *Mr Novak* that I had done…"

"In 'X is the Unknown Factor,' I played a youth who built computer robots and Dr. Baxter and the wonderful Anne Seymour played educators who wanted to evaluate my project. I talked with Dr. Baxter about computers and told him I had no talent or proclivity for them. I still don't. We both agreed they were the future like it or not."

Mirror image: lookalikes Dr. Baxter (left) and actor Dean Jagger peer at each other on *Mr. Novak*. Author's collection.

Even after the passage of fifty years, Macklin vividly recalls Baxter's charisma, engaging personality, and charm. "We also talked about acting and literature and Shakespeare. He was a delightful and erudite man."

Macklin also credits Baxter helping launch his acting career in a roundabout way. It seems that young David was one of those who watched the good doctor's *Shakespeare on TV* shows. As he says today:

Frank Baxter introduced me to Shakespeare over the air. I memorized Hamlet's "To be or not to be" at eleven. I performed it for Walter Eyer of "The Walter Eyer Theater" in my introductory interview audition at that age. Eyer seemed a bit "floored" by it. Years later he told me it was the youngest Hamlet and one of the best he had heard. I told Dr. Baxter about this and he got a kick out of it. I told him I didn't know what a "bare bodkin" was but just played it like I had to!"

Dr. Baxter is the obvious "draw" in this advertisement for the travel adventure show.
Author's Collection.

Baxter became much sought after as an onscreen host and narrator. In the 1957-58 season he was the host-narrator for *Telephone Time*, a dramatic anthology series. Baxter would be seen in the first five minutes of each episode, giving some background and setting the stage for the coming drama. The series was bankrolled by Bell Telephone, the same people that were behind the Bell Science series.

In the episode *The Vestris* for example, he is seen in a kind of nautically themed set, casually leaning on what looks like the piling on a pier. Baxter asks the audience if they had ever come into a room that was new to them, yet had the strange feeling they had been there before. The professor calls it "déjà vu," and explains that this upcoming sea tale is set aboard the bark *Vestris*, a ship bound for Boston in 1828.

He further explains that this story was originally written by Robert Dale Owen, a utopian reformer in the nineteenth century. This particular drama stars Boris Karloff, best known as Frankenstein's monster in the classic 1931 film. Baxter stayed with the series until it ended.

Dr. Baxter also hosted *Four Winds to Adventure*, a half-hour weekly series that showcased documentary travelogues that roamed throughout the globe. This 1966 series often displayed the work of filmmakers with their families. Baxter would set the stage as he did in *Telephone Time*, but also chat with the filmmakers and interview them. It's significant that Baxter's name is prominently displayed in ads promoting the show. He was the real star of the production, and his name alone would generate interest.

In his book *The PBS Companion: A History of Public Television*, David Stewart said, "If Frank Baxter had continued to concentrate on teaching *Shakespeare on TV*, his work would have been an interesting but isolated cultural event." True enough, but in that vast audience of millions there was one person who looked at the Shakespeare series with particular interest. His name was Frank Capra, a legendary Hollywood director who was branching out into television. He was planning a documentary of the sun, but in a new, radical, and completely innovative way.

Capra was so impressed by Baxter he asked the professor to co-star in *Our Mr. Sun*, the first of what was going to be a dozen science shows created for the general public. This was the start of the famous Bell Science series, a project that forever altered the way science and technology would be presented to the average person. And Dr. Frank Baxter of USC was going to be a key ingredient in *Our Mr. Sun*'s success or failure.

4

Besides Baxter and Bell: Other Early Science Shows on Television

The Bell Science series wasn't the only effort to help bridge the yawning chasm of misunderstanding that separated the general public from the scientific world. The first really major television science program was *The Johns Hopkins Science Review*, which premiered in 1948 and lasted until 1960.

The Johns Hopkins Science Review was the brainchild of Lynn Poole, director of public relations at Johns Hopkins University. He was not a scientist, but did have a background in art, and had ideas that might or might not appeal to the public. When Baltimore station WMAR-TV asked around for programming suggestions, Poole suggested a something that might showcase the university's faculty.

The station was happy to oblige, so the *Science Review* was born. It was carried by the Dumont Network, the nation's fourth network, until Dumont folded in 1955. Each early episode had a prologue in which a narrator says, "Here in the many laboratories Johns Hopkins scientists are constantly probing the still unknown secrets of science, which, when discovered, will be translated into benefits enjoyed by you, the people of America."

This point is repeated again and again in these shows, reinforcing the notion that scientists are not just ivory tower "eggheads" but dedicated people working for the common good. It's a point that all good science shows advocate even today. Certainly, Frank Capra would have approved of this approach. It's something that he did himself in the Bell Science programs.

Originally, Poole intended to stay behind the scenes as writer and producer, but eventually he started doing program introductions as a kind of quasi-host. Though he became popular with 1950s audiences, his early appearances seem dry and stilted, not unlike the "talking head" scientists he was introducing. Dressed in a business suit, Poole is balding and bespectacled, with a kind of oval face. Later, he became much more relaxed and personable, though he never quite achieved the level of Dr. Baxter's charisma.

In fairness, these programs were live, and production values were minimal. Poole was also handicapped by shoestring budgets, as Robert Yoder wrote in the August 21, 1954, issue of the *Saturday Evening Post* "They (the *Review* staffers) will borrow your grandfather's picture off the living room wall, a hair dryer from a beauty parlor, a uniform from a policeman. They borrowed $250,000 in diamonds for one show. When Poole realized he might be setting up the first televised jewel robbery, he also borrowed four Baltimore cops."

Later, the show was renamed *Johns Hopkins File 7*. Poole stayed on as host, and by 1959, near the end of the series, he was actually more interesting and lively than the scientists he was trying to showcase. By the late 1950s, he sported a completely shaved head look—a total baldness much like contemporary actor of the time, Yul Brynner.

Seen today, some of the episodes are dry and boring. "The Usefulness of Useless Knowledge," telecast on February 11, 1952, is mostly an interview with an academic whose delivery is sleep-inducing. The academic "talking head" rattles on, while Poole nods in agreement. A later "talking head" in the same episode doesn't help much when he discusses "congenital malformations."

Much more interesting was "The Master Glass Blower," which aired on October 8, 1951. Poole starts the show by displaying a glass-blown figurine of a swan. He mentions that most of us have seen such little souvenirs, especially at carnivals and circuses. The scene cuts to a carnival where a barker is loudly trying to sell some glass figurines. "Step right up, step right up!" he yells to a crowd "Take your pick of the bargain of the century!"

There's a little humor too, as when the barker says to a young observer, "Go away son—you're too young. Besides, you bother me." The barker is none other than John Astin, most famous for his portrayal of "Gomez" in the 1960s sitcom *The Addams Family.* At the time he was a twenty-one-year-old student at Johns Hopkins. Today, still active in his eighties, he teaches method acting at the university.

The rest of the show features one John Layman, a professional glassblower who made all the intricate test tubes and other glass equipment for the university labs. Poole talks to him as he is heating the raw, molten glass, and the exchange is fairly entertaining. This is an example of how Poole sometimes played audience surrogate, asking questions and making comments.

Though highly praised at the time, some of the shows still seem dry and pedantic when viewed today. Many of the scientists were stiff, awkward, and dull. However, *Johns Hopkins Science Review* many times rose to the occasion, delivering the goods in a fascinating and even innovative way. In one episode, they showed a live birth, and in another it talked frankly about breast examinations and mastectomies. Remember, this was the 1950s, when Lucille Ball in *I Love Lucy* could not say the word "pregnant."

On December 5, 1950, *Johns Hopkins Science Review* made some television history of its own, and in a very dramatic fashion. Dr. Russell Morgan of Johns Hopkins demonstrated a special new X-ray fluoroscope that could penetrate the human body in such a way as to watch the living organs "in action." An industrial worker named Carter had recently suffered an accident on the job and was in great pain. He was coughing blood and was in great distress.

It seems there were some metal shards lodged in his back—but were they operable? Dr. Morgan set up his X-ray machine, and started to examine the patient while viewers in twenty-six cities watched. The stage was set for the world's first demonstration of remote diagnosis. Two physicians, Dr. Paul C. Hodges in Chicago and Dr. Waldren Sennot in New York, were watching the same images on TV, just like the general public. They were also linked to the Johns Hopkins studio by telephone.

In a very dramatic exchange, Lynn Poole asked both doctors if they could see the images clearly. They both answered in the affirmative. The patient was instructed to take a deep breath, and the X-rays plainly showed the metal fragments in his body did not move. The doctors soon arrived at a diagnosis, and it was left to Dr. Morgan to let the patient know the results on live TV. "Mr. Carter," he announced, "I have good news for you. We can remove the foreign bodies surgically with relatively little difficulty."

Even at its peak, the *Johns Hopkins Science Review* only attracted perhaps 500,000 viewers, and funding was always a problem. Poole did use some showmanship to keep the audience's interest, like having women in bathing suits parade around, but he hesitated to go down the entertainment road. He didn't, as he once said, want to make "monkeys" out of prominent men of science, and by the same token he didn't want to turn the show into a "circus."

The next science show worthy of mention was the California Academy of Science's *Science in Action*, telecast from the Academy's home base in San Francisco. Launched in 1950, it continued with something like thirty episodes a year until 1966. The first host was zoologist Tom Groody, but he was soon replaced by Dr. Earl S. Herald, Curator of Marine Biology at the Steinhart Aquarium.

Herald was an ichthyologist, and fishes were his passion. After World War II service—he was a captain in the U.S Army's Sanitary Corps—he studied tuna and tropical fish. In fact, he discovered a new species of angelfish which was subsequently named in his honor, the *Centropyge Heraldi*. But Herald really hit his stride when he was hired as Curator of Marine Biology at the Steinhart. A workaholic, he labored twenty-four/seven to make the aquarium one of the best in the nation, and it paid off.

While no Frank Baxter, Herald had his own unique brand of genuine charm. At first, he was a bit dry and stilted, but over time he relaxed and became more comfortable in front of the camera. Before long, Herald genuinely began to enjoy himself, and the audience could sense this. He was a natural, and there was one trait he did share with Frank Baxter: an enthusiasm and love of knowledge.

Towards the end of the show he would smilingly announce, "Now don't go away, I'll be back with the Animal of the Week!" The critters could be most anything, from baby coyotes, to rabbits, or a spotted skunk. Herald was known for his humorous ad libs, and it was the animal segment that gave him the best opportunity to demonstrate his wit. When a possum peed on Herald's white lab coat, he remarked "It's a good thing television doesn't include smell-a-vision."

The occasional mishaps actually added to the fun. Once, a frightened bandicoot bit its handler, and on another occasion, a herpetologist named Chuck Shaw lost control of a giant reticulated python. The snake bit him, and, though he was bleeding, the show carried on as if nothing happened. Such incidents allowed Herald to come out with a quip or two, such as "If your picture is a bit out of focus, it's because the cameraman fainted!"

The half-hour show covered a wonderful variety of topics. Mentioning just a few show titles will give an idea of its scope. There was "Game

Birds," "Story of Cheese," "Scales, Feathers and Fur," "How TV Works," "Before the White Man," and "Bone Hunters." Sometimes there would be famous guests. A well-presented special show on submarines featured Fleet Admiral Chester W Nimitz, the CINPACFLT, that is, the Commander in Chief, United States Pacific Fleet in World War II.

Also appearing on that show was Vice Admiral Charles Lockwood, who commanded the U.S. Pacific Fleet in World War II. Both men are informative and gracious, and there are plenty of charts, pictures, and film clips to enhance the story line. As always, Herald plays "audience surrogate" with a touch of humor. When Admiral Lockwood comments that submarines feature the "best food in the Navy," Herald responds that he likes that—he likes food!

The show primarily aired in California, though kinescopes made it possible to have showings elsewhere. In the beginning of the show, there's a prologue featuring some anonymous scientists in white lab coats tinkering around a laboratory filled with test tubes and strange-looking apparatus. A message appears in capital letters, informing the viewers that "here in their world of utmost precision they toil endlessly for the advancement of knowledge."

Once again, the image is, like the Bell films, of the scientist as a hero, working for the benefit of humankind. The scientists are not long-haired atheists, or madmen trying to invent a new weapon of mass destruction. They are the people who are discovering things to make the world a better place to live in. This seems to be a common theme for the fifties science shows.

Science in Action finally left the air in 1966, a victim of rising production costs. As always, science programs were continually challenged for funding, a situation that continues for most PBS stations. Herald helped create many of *Science in Action*'s 626 episodes. He also authored ninety-three publications. One of his books, *Living Fishes of the World*, was a popular tome that was translated into eleven languages.

Herald formed a scuba diving research group and frequently went down to Baja California in pursuit of his lifelong passion: rare and exotic fish. Tragically, he drowned off Cabo San Lucas at the age of fifty-nine. Sad as it was, at least he died doing what he loved best. The California Academy of Sciences is the world class institution it is today thanks to men like Dr. Herald.

Walt Disney became interested in the "education as entertainment" concept as early as 1945. In that year, he wrote an article in *Public Opinion*

Quarterly that was entitled "Mickey as Professor." Disney strongly advocated "attitudes that are fundamentally educational although expressed in the manner of entertainment" through animation and dramatization. "Uncle Walt" did stop short of saying the new-fangled thing called TV would solve all of education's problems.

The man who created Mickey Mouse was also a conservationist who genuinely loved nature. He was worried about the inexorable advance of civilization in America's last frontiers— particularly Alaska. He hired the filmmakers Alfred and Elam Milotte to record all they could of the then-territory's wilderness. It was easy enough for the husband-and-wife team, since they owned a camera store in Alaska.

The original idea was to record everything—the Inuit (Eskimos), wildlife, towns, and whatever else they could think of. But once he had a look at the raw footage, Disney decided that nature, not man, should be the primary focus of his film. There was one segment that particularly caught his eye: seals that inhabited Alaska's wild shorelines

Walt asked the Milottes to expand the seal segment and emphasize the life cycle of the seals through the seasons. When the documentary was done, Disney had another problem on his hands: at the time his distributor was RKO-Radio Pictures, one of Hollywood's top movie studios. RKO balked at handling what came to be called *Seal Island* because it was new and innovative, and they were sure that audiences would not sit through a nature movie.

Disney, undeterred, asked a friend to run *Seal Island* at Pasadena's Crown Theater for a week in December 1948. The main purpose was to have the movie qualify for Academy Award consideration for the 1949 Oscars. *Seal Island* ended up winning the Best Documentary Oscar for that year. A triumphant Walt, Oscar in hand, walked into his brother Roy's office with the statuette in hand. "Here, Roy," Walt said, "Take this over to RKO and bang them over the head with it."

There were thirteen Disney nature science films in all, and they were in production until 1960. Walt Disney himself said that "We did not succumb to the alluring temptations to make villains or saints of the creatures portrayed in our films. We have maintained a sensitive regard for the wisdom of Nature's design and have attempted to hold a mirror to the out-of-doors rather than to interpret its functioning by man's standards."

Walt was sincere, and in the main he kept to those principles. However, nature films were new, untried creations, and he always remembered his target audiences were families with children. That led occasionally to missteps on his part. One example was *The Living Desert*, which premiered in 1953.

Much of *The Living Desert* is colorful, strikingly beautiful, and factual. But Disney had some sequences that overstepped the line from reality to manufactured fantasy. The most notorious is the mating dance of the scorpions, played out like a country square dance. Trick photography—like stopping the motion, reversing it, then moving it forward again—make the arachnids "dance" back and forth. While they cavort, deep-voiced narrator Winston Hibler calls out the square dance moves while hoe-down music plays. "Stingers up for the stingaree" he sings, "—now, watch out gal, you don't sting me!"

Disney was rightly called to task for such buffoonery, and he never did anything so blatant again. However, a segment in 1957's *White Wilderness* has since become an urban legend: the celebrated lemming mass suicide into the sea. The filmmakers, apparently without the knowledge of Disney or his company, bought—yes, bought—about a dozen lemmings from some Eskimos and took them to (landlocked) Alberta, Canada.

They took the lemmings to a snow-covered turntable to film them, with careful editing and camera work making it seem that the dozen were thousands. But what was really despicable was the way they filmed the dramatic ending. The filmmakers took the poor lemmings to cliff and either drove them off, or threw them off, a precipice.

The "ocean" was really a river, carefully photographed to make it appear to be the sea. Some stock footage of the real ocean was spliced in to reinforce the illusion. Of course, in real life lemmings do not have a "suicide instinct" to run and run and hurl themselves off of a cliff into the ocean. The whole thing was faked, and in the process created an urban legend.

But most of the time Disney was true to the facts, and his nature films—later labeled "True-Life Adventures"—hold up even today. Disney and the filmmakers were very exacting, and Walt would not stand for anything less than perfection. After all, these movies had his name on them. In general, for every 120,000 feet of film that was shot, only 30,000 was used. In an even more telling example, it took three years to film *The African Lion,* which was released in 1955. Only six percent of the footage was used in the final seventy-two-minute film.

Sometimes Walt was ahead of his time. *The Vanishing Prairie* (1954) faithfully chronicles the lives of the many animals who live there, particularly the American Bison (buffalo). One scene showed the birth of a baby buffalo, which caused the state of New York to actually *ban* the film.

Disney himself commented that the birth scene "would never have appeared on the screen if I believed it might offend an audience. It would be a shame if New York children had to believe the stork brings buffaloes too." New York reversed the ban. To obtain close up footage of the buffalo, filmmaker Tom McHugh covered himself and his camera with an old buffalo hide and crept in amidst a herd.

Walt also started producing what was called "True-Life Fantasies," which sounds like a contradiction, but made perfect sense in the Mickey Mouse kingdom. In these scenarios an animal—a real one—would be given a human name and put in a structured story. The first was *Perri*, the story of a little squirrel and the adventures in her life. Though slightly anthropomorphic, the story remained largely true to nature. Other titles followed, including *Sammi the Way Out Seal,* and *Charlie the Lonesome Cougar*

All of "Uncle Walt's" "True-Life" and "True Fantasies" ended up on TV in one form or another. Disney was smart enough to recognize the potential of television long before the other studios did. To most Hollywood studios the "little box" was a rival to be bested by Cinemascope, Technicolor, and wide-screen epic extravaganzas. Walt thought otherwise.

Disney's first television program began in 1954. Though it went through a number of name changes, essentially it was an anthology show that might feature cartoons one night, a Disney-produced drama another night, or a Disney theatrical film promo. But Walt was something of a visionary and had a genuine interest in science and technology, and this characteristic is often reflected in his shows.

He personally created what came to be called the "Carousel of Progress" for the New York World's Fair of 1964-65. The show was peopled by lifelike robots under a new process dubbed by Disney engineers "audio-animatronics." There are four tableaux: Spring, the turn of the century; Summer, 1920s; Autumn, 1940s; and the Present. In the summer 1920 sequence, for example, "John," the audio-animatronic host, sits in a house that features all the "latest" technology, like the radio, the electric iron, and other "marvels."

Walt also was fascinated by the exploration of space. This interest caused him to produce three episodes on his TV show that dealt with the subject: "Man in Space," "Man and the Moon," and "Mars and Beyond." It's important to note the first two shows were produced and aired in 1955, a full two years before the Russians inaugurated the space race by launching Sputnik, the world's first artificial satellite into orbit.

"Man and the Moon" can be considered typical of the trio. It's a very entertaining, imaginative mix of real science and fantasy—and Disney makes sure the two are well separated. A series of brilliantly conceived cartoons directed by Ward Kimball trace the history of man's association with the moon. Its place in popular culture is also explored, from Shakespeare to popular songs.

The focus then shifts to a more serious, though just as entertaining, discussion of the moon and the possibility of landing humans there. Animated sequences detail what a lunar flight and landing might look like, but this time in a more realistic fashion. Some of the animated images are actually quite beautiful.

The trio of space episodes was given an added authority by the inclusion of prominent rocket scientists like Dr. Werner von Braun and Dr. Willy Ley. Both men were later very important figures in the American space program at NASA (National Aeronautics and Space Administration). Von Braun was probably the most famous rocket scientist of the 1950s, who wrote a book idealistically entitled "*I Aim for the Stars.*" He had been involved in Hitler's V-2 rocket program, which caused jokesters to say his book should be retitled: "*I Aim for the Stars—and Often Hit London.*"

Nevertheless, Walt Disney and his creative teams should be honored for their work in making science and technology accessible to the general public. While well aware of pollution and other threats, Disney remained upbeat about science and humanity's future. The theme song of the *Carousel of Progress* pretty much sums up Disney's view of the future, with lyrics such as "There's a great big beautiful tomorrow, and tomorrow's just a dream away." For Walt, science and technology would be of vital importance in reaching that "beautiful tomorrow."

Oceanographer and undersea explorer Jacques-Yves Cousteau became famous as a result of his exposure on TV. In the 1940s, Cousteau was a key player in the development of an Aqua-Lung for diving, the forerunner of today's open circuit scuba technology. In 1951, he outfitted an old minesweeper and turned it into a research vessel to explore the sea. The ship, renamed *Calypso*, became almost as famous as its owner.

As Cousteau and his team explored the ocean depths aboard *Calypso,* much of their activity was filmed and shown in movie theaters. Oddly enough, Cousteau's American television debut was on *Omnibus*, the famous "cultural magazine" show. The Cousteau segment was entitled "Underwater Archaeology," and aired in January 1954. Captain Cousteau returned to *Omnibus* in 1956 to show his award-winning documentary *The Silent World.*

In 1961 Cousteau was honored when he was awarded the *National Geographic* Gold Medal at the White House with President John F. Kennedy. But it was in 1966 with his first television special "*The World of Jacques-Yves Cousteau*" that his small screen career took off in earnest. His special's rating was so high the American Broadcasting Company (ABC) took an interest.

The result was a series called *The Undersea World of Jacques Cousteau,* which began in 1968 and ran for eight seasons. Cousteau became something of a pop culture star in the sixties, and with the environmental movement in full swing, his message of ocean conservation was well received. The captain could speak good English, but it was with a delightful French accent. Cousteau would narrate his own shows, and I can still remember the distinctive voice, as "zees ess the beautiful mating dance of zee sea ottair (otter)."

Cousteau became such a pop hero that singer John Denver wrote *Calypso,* an award-winning song about his ocean explorations in 1975. The captain also founded the Cousteau Society, an international ocean conservation group. Some critics noted that Cousteau had no scientific degrees at all. In his younger days, he was an officer in the French navy. But he learned a lot during the course of a forty-year oceanic career, and probably no man has advanced the cause of understanding and protecting the ocean like Jacques-Yves Cousteau.

The National Geographic Society was also a major player in presenting science topics to the general public. In fact, some years they were just about the only game in town when it came to popular science. The Society was founded in 1888, and one of its leading developers was Alexander Graham Bell, the inventor of the telephone. Its stated mission was the "increase and diffusion of geographical knowledge."

The Society's magazine, with its distinctive yellow borders, became something that linked the generations. The writing was usually excellent, but it was the photography that set National Geographic Magazine apart from its competitors. Cliché or not, the magazine really did bring the world to the average American household. And there were other compensations as well: Generations of schoolboys caught their first glimpse of the naked female form via its pictures. This was due to a policy, even in the straight laced late Victorian period, not to censor photos of "native" women of the world.

In 1961, National Geographic established a television service and contacted Hollywood production groups to make and ultimately market documentary programs. The first National Geographic special aired in 1965. The Society wasn't above a little Hollywood hoopla to promote its fledgling effort. "Would you like to stand with Barry Bishop on the glittering top of Mt. Everest?" the ad copy breathlessly inquired. "Or visit Jane Goodall as she camps among wild chimpanzees? Beginning next month, you can enjoy such adventures in your own living room."

Primatologist Jane Goodall was the National Geographic's first certified "star." In *Miss Jane Goodall and the Wild Chimpanzees*, the society's second special, viewers were able to journey deep into the remote forests of Africa with a woman scientist who was not only intrepid and knowledgeable, but also young, blonde, and photogenic.

National Geographic urged its viewers share with Goodall "discoveries about animal behavior that have startled the scientific world." By watching the documentary they would gain "rare insight" into the "personal lives" of the animals. Sounds like a mammalian soap opera, and when Goodall saw the first "cut" of the film she was appalled. It was much too cutesy, and the "blatant anthropomorphizing" of the chimps did a disservice to both the animals and their human viewers. Luckily Goodall was able to persuade the filmmakers to make substantial revisions.

The National Geographic specials covered a lot of topics, and it generally did them well. Anthropology and archaeology seemed almost a specialty. *Dr. Leakey and the Dawn of Man,* for example, traced early hominid discoveries in Africa's Olduvai Gorge. Other shows might highlight a certain wilderness area like the Mohave Desert, or the efforts to preserve the grizzly bear. Whatever subject they tackled, the National Geographic specials were noted for their literate scripts, colorful location photography, and high production values.

No book on science TV would be complete without speaking of Don Herbert, a.k.a. "Mr. Wizard." Herbert portrayed a "neighborhood" scientist who demonstrated simple experiments for the benefit and education of children. He had almost no scientific education but managed to convey an air of knowledge and authority that made an impression on young minds. Unlike the Bell films, who wanted to reach all ages, Herbert's main audiences were in the eight to thirteen age range.

Watch Mr. Wizard premiered on NBC March 3, 1951, and was soon a big hit. To give the experiments more immediacy, Herbert used items that you might find in any home—milk bottles, coffee cans, knitting needles—to demonstrate such relatively abstract principles as gravity, magnetism, and oxidation. Mr. Wizard always used children as assistants, and in the very early shows he actually used a neighbor kid. Eventually, though, he employed child actors, who could memorize lines and show the proper—if staged—enthusiasm.

At first, the shows were live, but by the 1960s they were taped, in part to better coordinate classroom lesson plan prep. Over the years he became quite knowledgeable, and when Herbert tackled subjects that were beyond his range he wasn't afraid or too proud to bring in expert advice. In the 1960s, for example, New York University physicist Morris H Shamos was brought in as the show's behind-the-scenes science advisor, to ensure accuracy.

The Mr. Wizard persona was that of a friendly uncle, but there was no sugar coating in his programs. If a young assistant balked at a Bunsen burner flame, or other tool, for example, he'd remark "You have no confidence" or "it won't bite you!" Herbert was also a genius at marketing, authorizing Mr. Wizard Science Clubs across the country. Each local club would be named after a common element, for example the Minneapolis Aluminum Club.

These clubs were enormously popular, and provided a good tie in for the TV shows. There were over 5,000 Mr. Wizard Science Clubs by 1956, with thousands of children as eager members. Herbert starred in over 500 episodes before the series was suddenly cancelled in 1965. Unruffled, in 1966 he started a miniseries aimed at adults called *Experiment: The Story of a Scientific Search*. It focused on individual scientific endeavor, like being a geologist, and examined the challenges such men and women face.

In *Experiment,* Herbert was much less an "uncle" and much more the academic, wearing a suit and tie and presiding over experiments that were far from an average family's ability to copy at home. There were eight half-hour episodes in the series. But by 1971, parents groups and others were criticizing the networks for feeding kids a steady diet of mindless cartoons and endless hard-sell advertising. In response, NBC revived *Watch Mr. Wizard.* Alas, it was a brief renaissance; within one season, Herbert's show was once again cancelled.

Herbert's career continued well into the 1990s. With the proliferation of cable channels he didn't have to depend on the whims of the "big three" networks. From 1983 to 1990, Herbert landed a berth in the *Nickelodeon* channel with a show called *Mr. Wizard's World.* The 1980s incarnation of Mr. Wizard was very much like the shows he had hosted twenty and thirty years before. Most of the changes were minor; the programs were now in color instead of black-and-white.

Herbert also had to slightly adapt to modern times. His famous old egg in a bottle trick was changed to a water-filled balloon and soda-bottle one. As he explained, "most kids today have never seen a milk bottle." The series ended in 1990, but was seen in reruns until 2000. Herbert occasionally showed up on talk shows like *The Tonight Show with Johnny Carson*, and in 1992 he was the first guest on *Late Night With David Letterman.*

Don Herbert's career was winding down, but he wasn't quite through yet. In 1994, he did a series of fifteen-minute spots called *Teacher to Teacher with Mr. Wizard.* The shows highlighted individual elementary school teachers and their various projects.

Herbert died in 2007, just four weeks shy of his ninetieth birthday. William "Bill" Nye, who hosted a 1990s show called *Bill Nye :The Science Guy* wrote after Herbert's passing that "If any of you reading now have been surprised and happy to learn a few things about science with *Bill Nye: The Science Guy*, keep in mind it all started with Don Herbert."

The science shows of the 1950s and 1960s certainly made their mark in the history of science on television. But none of them, not even Mr. Wizard, was going to have the impact of the Bell Science films.

5
Genesis: The Bell Science Series Is Born

The Bell Science series had a long gestation, and many times during the course of its development it seemed as if the project would be stillborn. The American Telephone and Telegraph Company was a corporate giant in every sense of the word. In the 1950s, AT&T had 100 million phones in service—half the world total. Direct Distance Dialing (DDD) had been introduced in New Jersey in 1951 and was eventually extended to include overseas calls.

Most people were aware of AT&T through its "child," Bell Telephone. The Bell System was essentially a giant, government-sanctioned monopoly, and eventually acquired the infamous nickname "Ma Bell." But AT&T was more than just a corporation bent on maximizing profits; it also had a more benign, public-service side as well

AT&T President Cleo T. Craig was something of a visionary, and felt that science was going to be an ever-growing factor in the lives of everyday people. This was 1951, six years before the Russians launched Sputnik, but our cold war rivalry with the Soviets probably had a role in Craig's thinking as well. How could the United States maintain its leadership in technology if it lagged behind the Soviet Union?

It's not that Craig forgot the commercial aspects. Far from it. As Frank Capra once put it, "Mr. Craig insists that since science is what his company is selling, science is what the Bell System should sponsor." This made sense, because Bell Laboratories in New Jersey was one of the most prolific research and development facilities in the country. The Bell Labs helped develop radio astronomy, and created the transistor and the laser.

Craig conceived the idea of a series of popular science programs that would be bankrolled by his company. But he wanted something new, something different. Craig wanted his shows to entertain as well as inform, a novel approach for the time. And so, he turned to Hollywood—where else could he find the talent and expertise to fulfill his dream?

The AT&T president turned to the company's advertising agency, N.W. Ayer & Son, to set the project in motion. Specifically, Craig wanted the programs to air on television, to maximize their exposure to the greatest number of people. One always thinks of ad agencies as the ones who do the persuading, "pitching" ideas to their clients like a rapid-fire machine gun. In this case, it was just the opposite. Ayer expressed serious doubts about the idea.

The ad agency went and asked television experts, who heartily concurred. There would be *no* market in commercial television for science shows, however cleverly produced. The general public wanted to relax after a hard day at work, and preferred to watch mindless entertainment than to have to "think." Besides, science was too "deep" for the average person to understand, much less care about. Science was the realm of long haired "eggheads."

But Craig was not going to take "no" for an answer. He gave Ayer its marching orders: recruit the people needed to make a series of popular science shows for television. Donald Jones of Ayer's Radio-Television Division was given the task to start canvassing for Hollywood film people. Jones started phoning, but very few expressed any real interest. Most didn't care about science, and in those early days, television was a rival that was taking away movie patrons. One just didn't fraternize with the "enemy."

Nevertheless Jones persisted, and a film editor by the name of Gene Milford suggested director Frank Capra. In fact, Milford gave Capra his highest recommendation. "Frank's really bugged about science," Milford explained. "Whenever you work with Capra, you learn something." Convinced, Jones gave Capra a call just after the director had returned from India after attending an important International Film Festival held there.

In some of the interviews he gave later, Capra said he wasn't initially interested but quickly changed his mind. His film career was partly stalled, and his reputation was under a cloud—a cloud that was dissipating, but still a worrisome threat. Capra wasn't too sure about television, but the idea of breaking new ground in a new medium attracted him. He agreed to meet with President Craig in New York.

Frank Capra is one of a handful of directors—John Ford and Cecil B. DeMille are two others—that deserves the description "legendary." Of immigrant stock, he was born in Palermo, Italy, in 1897, and came to America when he was six. Capra was one of those people in the "Horatio Alger" mold, who by luck, pluck, and talent (and a good deal of hard work) became an American success story.

Capra was genuinely interested in science, and earned a degree in chemical engineering from the California Institute of Technology in 1918. He became a Hollywood writer and turned to directing. He reached his peak in the late 1930s, lensing a number of films that are considered classic to this day. His first really major hit was the romantic comedy *It Happened One Night* (1934) starring Clark Gable and Claudette Colbert. The film won a number of Oscars, including Best Director for Capra himself.

Other films followed, including the fantasy adventure *Lost Horizon* (1937). But he really hit his stride with a series of movies that celebrated the wisdom, decency, and honesty of the "common" man. In *Mr. Smith Goes to Washington* a young and inexperienced senator (James Stewart) successfully fights political corruption with the help of "grassroots" America. Though the plot is different, *Meet John Doe* (1941) also celebrates the courage and common sense of the average American.

Capra's films also glorify democracy and the American way of life. He soon had such name recognition that Americans would see a film because he directed it. His popularity was such that his name was always prominently displayed in a movie's opening credits. That's why his 1971 autobiography is entitled *The Name Above the Title.*

But times change, and so did American taste in films. After World War II, Capra's brand of unalloyed optimism seemed sentimental and naive to many critics, who labeled it "Capracorn." The world was different now, and the advent of the nuclear age seemed to usher in a new and sobering reality. Our rivalry with Communism and the Soviet Union brought with it a new responsibility and vision as the only superpower of the free world.

As if to underscore this new seriousness, *Film Noir* was now the rage, with its seamy urban settings and hardboiled detectives. Capra still worked but it was plain the glory years were over. In 1946 he made *It's a Wonderful Life*, which has since become a perennial Christmas classic. At the time, though, it was considered a commercial failure.

In addition to his career woes, Capra faced charges of political disloyalty. He was not alone, because in the years following 1945 there was an increasing fear of domestic Communist subversion, which later historians dubbed the "Red Scare." Loyalty boards combed through the records of federal employees, ferreting out all who had Communist party affiliation, or who had leftist associations in the past. Employees who were even the slightest bit tainted with the "disease" of leftism were summarily dismissed.

Then the U.S. House of Representatives got into the act, establishing the House Un-American Activities Committee (HUAC) to investigate Communist activities at home. HUAC took particular delight in investigating the Hollywood film industry. Movie stars, directors, and producers like Walt Disney were called to testify. Unfortunately, during the Great Depression, when capitalism seemed like it was failing, some Hollywood types had flirted with Communism or socialism. It would come back to haunt them.

The fog of Communist paranoia was so thick it refused to be burned away by the sunlight of truth. Few people were willing to criticize HUAC or any similar organizations. Singer Frank Sinatra, doing it his way, was one of the handful who did say something. As he put it, "If someone stood up for the underdog, will they call you a Commie? Are they going to scare us into silence?" Unfortunately, for many the answer was "yes."

Frank Capra was one of those who found himself tangled up in a web of suspicion. When the Korean War broke out in 1950, Capra patriotically volunteered for active duty. He was in his early fifties, well past military age, but he had made a brilliant series of anti-fascist propaganda films called *Why We Fight* in World War II. There seemed to be no reason he couldn't not have served in some similar capacity once again.

The U.S. Army turned him down, and he didn't know the reason until many months later. Privately G-2 (Army Intelligence) had reviewed his record and found him "possibly subversive." In the heated political climate of the time, that meant a "Red"—out and out Communist—or a "Pinko"—a Communist sympathizer or fellow traveler. A label like that could destroy a career.

At first, all seemed normal. The Defense Department asked Capra to serve on one of their think tanks: Project VISTA. Happy to serve, the director readily agreed. But then, disaster struck. Capra reported to VISTA but was stopped at the gate by a security guard. He was told to report immediately to a Colonel Miller's office on the property.

As soon as the director entered, the colonel rose from his seat. This meant a formal pronouncement of some kind. "Mr. Capra," Colonel Miller began, "I'm sorry to have to do this, but I am forced to relieve you of your identification card and all pertinent papers in your briefcase."

Capra was shocked almost beyond words. He was asked if he had received the letter that refused his security clearance. The director said no, so he was given a copy of what had been sent to him. Even as his eyes scanned the lines, he could feel a perceptive change in the colonel's attitude towards him. As Capra later recalled in his autobiography, "He fixed me with a cold look. There it was. Suddenly. Suspicion gulf a mile wide. His eyes moved away from the contemptible. Authority had spoken. Instant conviction. Traitor."

The director later admitted that the drive from the VISTA offices was a "Kafkaesque nightmare. I must have pulled off the highway a dozen times, to think, pound my head." On December 14, the other shoe dropped. Capra received a curt letter from the Army-Navy-Air Force Personnel Security Board (ANAFPSB) that he was *not* cleared for VISTA. It listed a number of charges, including association with "left wing" writers. He was given ten days to compose a reply.

It was December 18, 1951. For the next ten days, including Christmas, Capra and his secretary Chester Sticht composed a rebuttal that "renounced and denounced" all the charges laid at his door. The missive was accompanied by character references from prominent people, including fellow director John Ford. The final draft ran to some 220 pages of facts and figures. Capra then mailed it, trusting to God and the U.S. Postal Service.

For twelve days, "twelve days of agonizing silence," Capra waited and wondered. The wait ended when he received a long distance phone call from a Turner Sheldon of the U.S. State Department. Sheldon was very upbeat, and asked Capra if he could go to India for a film festival, which was really a thinly disguised Communist propaganda showcase. The subcontinent was an important third world, "nonaligned" country. It would be best, Sheldon explained, if Capra showed up as a kind of U.S. counterbalance.

"India?" Capra roared into the phone receiver, "I don't want to go to India. I won't even go to Glendale until you government jerks clear my good name and apologize. You hear?" Sheldon, unruffled, was sweetness and light and assured Capra all would be well. The director was seething. "The military's Board of Torquemadas kicks me out of VISTA because I'm supposed to be a Commie lover; the State Department begs me to go to India because I'm supposed to be a Commie fighter."

But Capra's prospects brightened in the new year. On January 14, 1952, he received a notification that the infamous ANAFPSB had cleared him. "Suspended life," as he recalled "began breathing again." He went to India, and the trip was deemed a success. Now back home, he was off to New York to meet with AT&T President Craig. To sweeten the deal, AT&T would pay all travel expenses, and there would be a consultation fee in the offing as well.

Capra soon found himself in Cleo Craig's spacious office. A ruddy faced, sandy haired man, Craig seemed an almost stereotyped Scot, evoking memories of "glengarries and bagpipes" in the director's mind. The meeting also included Harry Batten, head of the Ayer Agency; Jim Hanna, head of Ayer's Radio and Television Division; and Don Jones, the Ayer man who had initially made contact with Capra.

Hanna came to the heart of the matter.

> *Frank, you know as well as I do that science documentaries are a dime a dozen because they're all dull as dishwater. But Mr. Craig insists that since science is what his company is selling, science is what the Bell System should sponsor. But as an agency that has to try and get the biggest audience for the client's buck, we recommend against a science show for AT and T's first entry into the TV rat race. And we're sort of hoping you'd agree with us.*

Privately, Capra still wasn't sure he wanted to be involved, but his love of science made him come to its defense. "If the discovery of continents, planets, and man himself is dull, then men like Galileo, Newton, Magellan, Freud, Einstein, Fleming, and Alexander Graham Bell led lives as dull as dishwater. Which is not so."

Gratified by Capra's answer, Craig said "Let's go to lunch." Capra couldn't help being dazzled: the son of Salvatore Capra, poor Italian immigrant from Sicily, was hobnobbing with one of the most powerful men of corporate America. Capra told Craig he'd produce a pilot film for the project. He was impressed by one of the things the AT&T president had told him when the two men were alone: "Anything second best is not acceptable to the Bell system. Please remember that, Mr. Capra." That was an article of faith in Capra's own career—that coming in second is the "same as coming in last."

In a later interview with the Catholic periodical *Sign,* Capra revealed other reasons for wanting to do the series, in spite of his initial reluctance. The director recalled his own disappointment that his sons were not interested in pursuing a science career. Most people didn't think too highly of science or scientists. They were the ones, at least in the public's eye, who were tampering with nature, not exploring it. The "eggheads" had split the atom, and in so doing created the most horrible weapon in history, a weapon capable of destroying the whole earth.

Capra felt even scientists felt the public's animosity. The director said that "scientists themselves complain they are thought of as strange ducks—men who speak a queer jargon, write only in formulas, and are probably up to no good." Frank Capra was determined to change that low opinion through the Bell films.

He also wanted to dispel the myth—as he saw it—that all scientists were by definition agnostics at best and atheists at worst. Capra attended Cal Tech as a young man, and it was there where he supposedly reconciled his strong Catholic beliefs and his love of science and scientific pursuits. In his view there was no conflict between science and religion—none whatsoever.

In fact, Capra drew inspiration from one of his old professors at Cal Tech, physicist Robert A. Millikan. "I noticed," Capra said in the *Sign* interview, "that Millikan never gave a lecture without mentioning God. His lectures were truly exciting, inspiring accounts of what the mind of man can do. He saw no conflict between science and religion."

Millikan was probably the inspiration for the Dr. Research character, though Frank Baxter's own personality and love of learning was another strong element in the overall "mix." Dr. Research is kindly, affable, avuncular, and very knowledgeable. By the same token, Research is ever mindful of God and God's place in creating the world and all the creatures in it. He's not afraid to mention the Almighty in the same breath as he mentions scientific concepts.

What about the subject matter? Capra and Don Jones decided it would be the sun. And why not? In Capra's view it "was chock full of interest to every man, woman, and child in the world."

It was decided that there would be two or three screen treatments of the first show, each penned by a different writer. That way, Ayer and Bell could pick which version they thought was best and go from there. Capra would select two writers, and eventually he authored a third version for Ayer's consideration. The two men Capra selected to submit screenplays were German American scientist Willy Ley and the writer-philosopher Aldous Huxley.

Willy Ley was literally a rocket scientist, a man who had been interested in space travel while still a youth in 1920s Germany. Rocketry was all the rage in pre-Nazi Germany, and Ley was its chief propagandist and populizer. He wrote *Die Farhrt ins Weltall* (Travel in Outer Space) and was one of the first members of the German Spaceflight society. His enthusiasm was boundless, and he wrote numerous articles on rockets for both German and foreign consumption.

But the rise of Hitler left him depressed and anxious. A fervent anti-Nazi, he fled Germany in 1935 and came to the United States, where he lived the rest of his life. After World War II he was, along with Werner von Braun, probably the most famous German rocket scientist in America.

Although English was not his native language, he was a very good popular-science writer. His *Conquest of Space* and *Conquest of the Moon* are considered classics to this day. Capra invited Ley to his suite in the Sherry Netherlands Hotel in New York, explaining the Bell project over coffee. Ley agreed to write a screen treatment, and in return he would get a $900 fee.

Ley went to work, but the finished product left everyone, Capra included, dissatisfied. For all his brilliance on the printed page, he turned in a rather limp, run-of-the-mill treatment that was decidedly pedestrian. Strike one, so to speak, but Capra next turned to Aldous Huxley. He came back to California and discussed the issue with Huxley in the latter's King's Road home.

It's hard to pigeonhole Huxley because he was a man of many talents. Poet, philosopher, travel writer, pessimistic futurist, novelist, Hollywood screenwriter. The British-born Huxley was acknowledged as one of the great intellectuals of the twentieth century, but he was also an eccentric who was fascinated by spiritualism and hallucinogenic drugs. In fact, he famously asked for a shot of LSD shortly before he passed away in 1963.

Huxley accepted the Bell assignment for a fee of $5,000. Even if they ultimately rejected his treatment, Bell got their money's worth out of the writer. As Huxley confessed to his brother Julian in January 1953, he had to read 100 percent of the scientific material to leave out ninety-nine percent for a television audience. Huxley did indeed work hard but seemed to be oblivious as to what Bell really wanted. They wanted an upbeat, positive view of science and technology that would be suitable for public consumption.

Huxley seemed to want to use the Bell show to advance his own pessimistic agenda. Would he be a Jeremiah, a thunderous voice urging people to turn away from the false gods of science and civilization, or would he be a Cassandra, doomed to have his warnings fall on deaf ears? He plainly hoped to be the former, not the latter, and he poured all his energy into making a very pessimistic view of the world.

In Huxley's script, the world is fast depleting its natural resources, resources that have taken mother earth millions of years to accumulate. There's only so much oil, coal, forests, and precious metals in the world, and humans are squandering them at an accelerated rate. In the narration, Huxley's script warns "It isn't too difficult to be rich, which you're spending your capital. But when there isn't any more capital, what then?" The words are said over graphic images of famine in India, eroded soil in Mexico, and devastated forests.

Huxley also denounced atomic energy, admitting that it might be a new source of power, but that radiation might produce genetic damage and deformation. The British writer also raised doubts about defense spending, and the arms race that seemed to be developing. A noted pacifist, Huxley couldn't resist taking a swipe at Cold War policies.

The script that Huxley presented was rejected, but it had a number of elements that Capra later incorporated in the finished product. The Sun was the principle narrator in the Huxley version, a kind of anthropomorphic hydrogen-helium gas bag that tells its own story. There was also an element of spirituality in the Huxley version. One scripted scene shows Saint Francis of Assisi on an Umbrian hilltop reciting his Canticle of the Sun.

Capra knew both Ley's and Huxley's treatments were going to be rejected, and he was right. As a backup, Capra wrote his own version, more "theatrical" than the others. He drew his inspiration from Dr. Donald Menzel's popular science book *Our Sun,* published in 1949. Capra injected a little showmanship, a little Hollywood pizazz, to the proceedings. "I worked out a showman's treatment," he later noted, "in a play form, all in dialogue, with two principal live characters, Dr. Research and a Fiction Writer, and four principle animated cartoon characters, Mr. Sun, Father Time, Thermo the Magician, and Chloro Phyll."

The director mailed all three versions to Ayer, together with a note that said he considered his obligations to the project done. And this was not grandstanding or fishing for more money. He had doubts about actually producing the show. Within a few days, Jones called Capra long distance—still a relative novelty in 1952—and told the director the "showman's" version had won hands down. Would he come to New York again to meet with the Scientific Advisory Committee?

Capra hopped on a plane, and before long he was in a room with the Committee—an assembly of some of the most brilliant scientists of the day. AT&T president Craig was on hand as well, to introduce the distinguished members to the director.

There was George W. Beadle (biology), Caltech; Dr. John Z. Bowers (medicine), University of Wisconsin; Professor Paul R Burkholder (bacteriology), Brooklyn Botanical Garden; Professor Farrington Daniels (chemistry), University of Wisconsin; Dr. Maurice Ewing (geophysics), Columbia University; Dean George R. Harrison (physics), M.I.T; Dr. Clyde Kluckholn (anthropology), Harvard; Dr. Warren Weaver (mathematics), Rockefeller Foundation; and Dr. Ralph Brown (engineering), Vice president in charge of the Bell Research Laboratories.

There's no doubt that the Mr. Sun project was attractive to Frank Capra, at least on a certain level. But in spite of his recent troubles he was a Hollywood director with a solid line of hits to his credit. Television was growing in popularity but Hollywood still considered it an "anemic" little medium, scarcely worthy of consideration. If Capra produced this science show, would he lose face? Would it seem like an act of desperation after his postwar slump?

Besides, Capra, a devout if somewhat less than orthodox Catholic, had very firm views on science and religion. In fact, in his mind they were linked together. Capra still wasn't sure about taking this job, and part of him wanted to get off the hook. Well, he felt now was the time to spill the beans about his religious beliefs, and how they related to the study of nature and nature's laws.

With his dark brown eyes boring into his listeners, Capra announced:

> *Gentlemen, I'm not our man. You gentlemen are scientists. A physical fact is your truth, your bible, your discipline. Well, to me, a physical fact is boring unless—it is illuminated by a touch of the Eternal. So, you see, if I make a science film I will have to say that scientific research is just another expression of the Holy Spirit that works in all men. Furthermore, I will say that science, in essence, is just another facet of man's quest for God.*

Capra expected his little speech to be anathema to the assembled scientists. Few, if any, of them thought of their work as seeking the Almighty. Nevertheless, the committee was full of praise for Capra's concepts, and thought his idea of having animated characters explain complex ideas nothing short of brilliant. Why, it was like Aesop's Fables! Capra couldn't figure them out. Why didn't they reject him?

Finally Dean Harrison, the M.I.T. man, spoke up.

> *Frank Capra, scientists feel there is a gulf, a widening gulf, between science on one side and Mr. Average Citizen on the other. We have become members of this advisory committee in the hope that we can help you build a bridge across that gap. An artistic bridge, a spiritual bridge if you will, that will open up a two-way traffic of understanding between scientists and other human beings. You build such a bridge, Frank Capra, and you will accomplish much for yourself and the telephone company, and much more for the nation and perhaps for the world.*

After that heartfelt speech, Capra tried to make a few more feeble attempts to get off the hook, but he knew he was "caught." The director signed the first of several contracts—the later ones were revisions of the original, mainly because of production delays and changes in the scripts. Capra agreed to produce thirteen shows for a series that would begin on 15 December, 1952 and would continue at four-month intervals through 1953.

There seemed to be a brief "honeymoon" period between Frank Capra and the group of scientific advisors. Nevertheless, Capra wasn't all that happy about these professorial "eggheads" looking over his shoulders. He had a vision of what he wanted to do, and he didn't want too many "cooks" to spoil his creative broth. As he once remarked, he felt the scientists were well paid to "act as a front, a shock absorber, an awning to keep pigeon droppings off Mother Bell's virginal image."

Before being given the green light to go into production, scripts would have to be approved by the Advisory Board as well as project specialists that might come in at any time. Capra, however, did retain the right to do the final cut. In the end, the production schedule proved hopelessly naïve and unrealistic. It was going to take almost four years for *Our Mr. Sun* to be aired on television, and if you want to count the production time from Capra's first meeting with Craig, it was almost five years.

The first thing Capra did was to put a heavy veil of secrecy around the production of *Our Mr. Sun.* He insisted that he would be known as "Mr. X" in all Ayer correspondence, lest the word of his involvement leak out prematurely. Don Jones, Capra's Ayer liaison, felt the director simply didn't want to let the cat out of the bag too early because of the "experimental" nature of the production. Still, somehow a few tidbits of information did manage to escape Capra's cloak-and-dagger scrutiny. *Daily Variety* columnist Sheila Graham reported in April 1952 that Capra was working on a television series about "rockets, atom bombs, n' (sic) things." The formal announcement of his involvement came in October of that same year.

Capra began commuting from his Red Mountain Ranch in northern San Diego County to his Frank Capra Productions offices at 9100 Sunset Boulevard, Los Angeles. For convenience, so he wouldn't have to drive back and forth, he would often stay at a suite of rooms at the Beverly Wilshire Hotel. He assembled a good team around him, including editor Frank Keller, Chet Sticht, and Dolores Waddell.

Though he later became disillusioned by Ayer and (to him) its seemingly penny-pinching, tight fisted ways, the first few years were enjoyable. Looking back in 1971, Capra recalled that "I was having the time of my life on our Professor Santa's Workshop, as Keller called it." In spite of later difficulties, nothing could erase the warm memories. As he put it, "Alone with typewriter and Movieola, the magic of film, and the freedom to create, the days passed like hours the months like days."

By the 1970s, some of the later bitterness dissipated, because Capra admitted: "The years 1952-56 were as productive and packed with achievement as the war years of 1941-1945." This was a reference to his celebrated *Why We Fight* series, which was so admired, British Prime Minister Winston Churchill felt the films should be required viewing for all of the Allied armed forces. But in this case, to Capra's obvious delight, "I was not revealing the ugly facts of war, but the awe and wonder and fascination of nature to youngsters from eight to eighty."

Capra's comments also confirm an often-forgotten fact: The science shows like *Our Mr. Sun* were always intended for general audiences not just children. Today, while the Bell films are often rightly praised, sometimes they are described as "kids' shows." This is largely because they feature animation, and in later years they were indeed screened for elementary, high school, and even some college students. But they were never "kids' shows" strictly speaking.

But there were dark clouds in Capra's optimistic valley of sunshine, clouds that became stormy on many occasions. The director worked hard on the script, but was increasingly at loggerheads with the scientists on the Advisory Board. Perhaps his greatest critic was Dr. Donald Menzel, whose book *Our Sun* provided the foundations for Capra's first film.

Menzel certainly had the academic credentials to be an advisor, even apart from his popular book. He was a noted astronomer, and head of the Astronomy Department at Harvard University. At first, Menzel seemed sympathetic to Capra's ideas, at least when the scientist joined the project early in 1953. After all, his *Our Sun* was a conscious attempt to educate the general public in an entertaining way.

But Menzel simply could not see any trace of God in scientific discoveries or of any designer in the laws of nature. Capra's approach definitely linked religion and science, which rubbed Menzel the wrong way. The Ayer man Don Jones tried to be a conciliator between the two men, saying to the astronomer, "The script is still in its formative stages, and we welcome your comment on its tone."

Matters came to a head when Metzel took issue with the end of the film. As the scene unfolds, "Dr. Research" offers a kind of prayer of gratitude to an anthropomorphic sun. "We used to worship you as an unknown, whimsical god, but now we know you better… Thanks for our daily bread.. and the fruits and flowers you grow for us. Thanks for feeding our animals. And thanks especially for the glory and beauty of your 'good mornings'—may God keep you shining forever."

Menzel was scathing in his remarks, saying that a "whimsical" sun was all wrong. The final sentences, he complained, "bordered on the pantheistic." Menzel also sarcastically suggested that the script "quit repeating the word 'thanks' at least after the first appearance of it. And I suggest that you say food instead of 'daily bread.' Otherwise it looks too much like the Lord's Prayer directed to the sun."

The astronomer also ridiculed the very last sentence. "Even God lives by the laws of nature, one of which is that there is no such thing as perpetual motion. The sun will not keep shining forever and cannot keep shining forever, and this last sequence I find extremely grating, although I understand the basic reasons for it."

Menzel finally concluded that if religion must be brought into this somehow or another, Ayer should hire someone "who is skilled in that particular field." The implication was, of course, that Capra was *not* that man. When Capra read Menzel's comments he was livid, and he "erupted" like a volcanic explosion from Sicily's Mount Aetna. Eventually he calmed down, and he wrote a measured response to the astronomer's views.

Capra acknowledged that Bell had an obligation to the scientific community, but since the company goes into millions of American homes with their products, "they have a greater obligation to the American people… to acknowledge that all good things come from God—even science."

But Capra also drew a line in the sand. He might bend a little, but he would not retreat one inch from his basic philosophy. The director couldn't help taking a few pot shots at Menzel and his fellow "eggheads." He didn't name Menzel directly, but if the shoe fit, let the astronomer/ astrophysicist wear it. In Capra's view the objective of the program "is to show that science is an integral part of the lives of everybody, not just an esoteric pursuit of a select few in ivory towers who are indifferent to the laws of God and country."

This tirade seems to call scientists atheists, the very label these shows were meant to dismiss. Also, the director seems to imply that by ignoring the laws of "country" they were somehow unpatriotic. But Capra, now that the knife was in, didn't mind twisting it a little. "I have no intention," he declared, "Of glorifying a race of Brahmins who consider faith in God and the moral laws as fairy tales for the untouchables."

One can almost imagine Menzel wincing at these words. Capra made other points: Referring to God's laws is not ridiculous or out of place; and scientists should get off their high horse—they are not better than the average man. As far as Capra was concerned, "This is just to point out that scientists are doing God's work, just as surely as carpenters, teachers, or social workers."

Menzel threw in the towel after this exchange, at least to the extent of accepting, however reluctantly, that Capra's religious notions would pepper the finished film. The astronomer might have been silenced, at least for the time being, but other scientists raised other objections. Astronomer Otto Struve didn't like the idea of cartoon characters in the show, especially making the hero of the story, Mr. Sun, have human emotions and responses.

Yet another astronomer, Walter Orr Roberts, wanted to have the script change the public's notion of what a scientist really was. Roberts declared that he was neither "long hair or atheist." He was referring to a portion of the script in which Mr. Sun asks why the public distrusts scientists. Dr. Research answers that many people think scientists are "long hair atheists who make atom bombs." But in general, Roberts was delighted with the script, and was one of those "eggheads" who actually liked the God references.

Some years later, the third Bell Science entry, *The Strange Case of the Cosmic Rays*, was even going to stir even more controversy between Capra and his scientific detractors. There was a long scene in one of the early script treatments that illustrated a point by referring to the Lord's Prayer as being inscribed on the head of a pin. The script was trying to establish how very tiny molecules were, but Capra's attempt to once again introduce religion was again resented by the advisory board

But the scientists really balked at another line in the *Cosmic Rays* script, which defined science as "the art of discovering what God created." There were loud yelps of protest on that one. In fact, this time the majority of the board strongly objected. Capra's script assertions "unnecessarily mixes science and religion." They also wanted Capra to know that "science is not at present concerned with God's creations in the physical world, only the physical."

The advisory board accepted the idea of literary marionettes—Poe, Dickens, and Dostoyevsky—being part of the film's scientific quest. But they didn't like the ending: "...because of the sneering voice of Edgar, the implication that science has no motive, and the sad ending." The ending was changed to be much more upbeat, but the acrimonious debates on the scripts continued until Capra finally left the series.

During the *Our Mr. Sun* battles, Menzel was bloodied but not completely defeated. He still took pot shots at Capra's script and won a few small battles, even if he ultimately lost the "war." Early in 1956, with the production in full gear, Menzel somehow persuaded Capra to drop the reference to the birth of Christ as a legitimate date. But such small victories were few and far between.

Don Jones started to get a little nervous about those religious references, but in a different way. Was Capra's "preaching" theologically sound? To make sure, he sent copies of the script to experts at the Jewish Theological Seminary, the Catholic Archdiocese of New York, and the Union Theological Seminary. Capra's script passed with flying colors.

Our Mr. Sun was essentially finished by 1954, but AT&T seemed hesitant to air it. For all their initial enthusiasm, they started to get what amounted to corporate cold feet. How would the public respond? And what about AT&T's image in the minds of the American public? In the fifties, most people seem to have accepted the fact that AT&T and its child "Ma Bell" were a gigantic monopoly. Nevertheless, the company wanted to refurbish its image as well as promote science.

Leaving nothing to chance, AT&T decided to preview *Our Mr. Sun* to select audiences, starting about eighteen months before the show actually aired on NBC. Test audiences would view the film, and then be handed out a questionnaire to fill out. The form would ask them what they liked about it, what they disliked about it, would they want to see it again, and so on.

The company seems to have targeted specific groups within the American public: students, teachers, housewives, and AT&T employees. At least five different public reaction tests were conducted. On June 15,

1954, in Philadelphia, the company assembled forty-four science teachers and 142 students for a sneak preview. *Our Mr. Sun* was presented on film, not on television.

The Philadelphia students' reactions were particularly gratifying to the telephone company. After the screening, sixty-one percent of the students said they "most respected" the fields of science and medicine. A majority of high school students also recorded that they now wanted to take "more science courses," though this desire was less in the college students that viewed the film.

Interestingly, many students declared on their questionnaires watching *Our Mr. Sun* made them want to take more courses in history, French, and English. This seemed to indicate an appreciation for knowledge in general, not just in science alone.

On June 19, 1955, *Our Mr. Sun* was shown at Riverside, California, right after the regular feature, Doris Day's *Love Me or Leave Me*, was screened. The reaction that the audience had was generally favorable and most said they wouldn't mind seeing it again. A few actually didn't like the religious references, noting the "incompatible philosophy of science and religion." A few others decried what they considered "gimmicks" in the film, whatever that meant. It is significant to note that a large majority favored films of this sort.

It seems that the people who objected to the religious references were in the minority in the test screenings and even more of a minority when *Our Mr. Sun* was nationally televised. When the film was finally shown on television, only about one per cent of the audience correspondents objected to the religious aspect of *Our Mr. Sun*.

After the telecast in 1956, AT&T followed up with phone surveys of 3,006 households in Toronto, San Francisco, New York, and Chicago. The company, seemingly obsessed with how the public perceived their show, also carefully monitored the fan mail that was received after the first broadcast. The 1950s audiences—with some exceptions—actually liked the religious references that Capra liberally sprinkled through the script. Curiously, sixteen of the forty-four teachers in the Philadelphia showing "most liked" the way *Our Mr. Sun* combined "scientific accuracy...with religious spirit."

Some of the fan mail that Capra received showed many people wanted science and God to join forces with one another—presumably to battle the atheists of the Communist world. One ardent fan gushed that *Our Mr. Sun* seemed nothing less than "a message from God, himself." To another fan, Capra was showing the best way of how to utilize this "God-given medium."

It's also interesting to note in all those screenings and telephone surveys the majority of the people used and contacted were women. In one telephone survey, for example, seventy-seven percent of the people contacted were female. On reflection, it makes perfect sense. The men and women who worked during the day would not have the time or opportunity to look at screenings and fill out questionnaires. "Housewives" would have the time to do such things.

AT&T actually had a survey category, "housewives." This is certainly something that would not be considered today. In the 1950s, a detached home in suburbia was the ideal, with men going off to work and women staying home to attend to domestic chores. There's even a segment in *Our Mr. Sun* featuring a cartoon lady dressed up in typical '50s style. "The sunlight on this lady's parasol," Eddie Albert glibly informs us, "has one and a half horse power constantly pouring on it—enough to run her washing machine, sewing machine, refrigerator, and vacuum cleaner."

After the test screenings were over with, AT&T finally gave the green light for *Our Mr. Sun* to be telecast just before Thanksgiving 1956. The preliminary months of argument and doubt were over. The Bell Science Capra years were about to begin.

6
The Capra Years

The four Bell Science programs that Frank Capra produced were *Our Mr. Sun* (1956) *Hemo the Magnificent* (1957) *The Strange Case of the Cosmic Rays* (1957) and *Meteora: The Unchained Goddess* (1958). All four had their challenges, but *Our Mr. Sun* took the longest to complete and air on television. This was due to the time-consuming nature of the animation, and also the various script wrangles that Capra had to endure over a period of about four years. Capra was also trying something new: namely, trying to present fairly detailed scientific knowledge as both accessible and entertaining. It was a painstaking process.

The animation sequences were going to be a major component in *Mr. Sun*—the glue that would hold the narrative together. Capra chose U.P.A. (United Productions of America) for this all-important task. Today, U.P.A. is probably best remembered for its Mr. Magoo cartoons. Director of Animation, William Hurtz, like so many of his generation, had roots as a Disney artist. But after World War II he left "Uncle Walt" to help found U.P.A. A brilliant graphic designer, he helped popularize a minimalist style that was light years away from the Disney offerings.

Minimalist animation of the 1950s was radical for its time, but it breathed new life into the art. Hurtz used abstract art, symbolism, and limited movement. Everything was stylized, simplified, with understated motion, clean lines, and striking color. There was no attempt, as in Disney's films, to realistically duplicate nature. Viewers were transported to a vibrant new world, a world that was minimal enough to let their imaginations fill in the blanks.

In 1951, Hurtz had been the layout artist for *Gerald McBoing-Boing*, a landmark film that introduced the minimalist style and won an Academy Award in the process. In fact, Hurtz himself received an Academy Award nomination for *The Unicorn in the Garden* in 1953, just about the time he was involved with *Mr. Sun*. Capra could not have chosen a better man for the first Bell feature.

Hurtz's first task was to design an effective Mr. Sun. As Hurtz recalled years later, "Tony Rivera had designed a knob-nosed sun—a more conventional Mr. Magoo type of sun. I gave it sharper features, with lines radiating from the eyes. The idea was to have radiant expressions: He emitted the lines of rays when he was smiling or upset, which I think worked very effectively."

Other characters included the goddess Dawn, Father Time, Thermo the Magician, and Chloro Phyll. Each had its own personality and role to play in the unfolding story. All of the cartoon sequences showed Hurtz's particular genius for making memorable cartoon characters. The anthropomorphic Sun and his companions were going to appear on a "magic screen," while "Dr. Research" would control a second screen reserved only for facts.

There were a lot of "straight" illustrations in the "fact screen," though presented in a cartoon or sketch format. These fact graphics are a highlight of the show, as when Dr. Research is talking about the birth of the sun. Gravitational forces are shown as two giant hands compressing the solar disc. This is a classic Capra/Bell technique: striking visuals that create easy-to-understand and easy-to-remember facts.

There were going to be only two live actors in *Our Mr. Sun*, so casting the right people was one of Capra's primary concerns. The two were "Dr. Research," and a "Fiction Writer." Perhaps Dr. Research was more important than his onscreen colleague, because he was going to be the primary source for information and facts. If he came off as dull, dry, or pedantic, Capra's whole concept of popularizing science would collapse like a house of cards.

Capra wanted to get someone with genuine charisma, someone who would lay to rest the public's notion of scientists as either long haired atheists or boring, cerebral academics hiding in their ivory towers, isolated from the rest of humanity. Then Capra remembered Frank Baxter and his wonderful *Shakespeare on TV* lectures. The director had been a viewer himself and was spellbound by Baxter's uncanny ability to teach in a warm and engaging manner.

Frank Capra directing Dr. Baxter and Eddie Albert on the set of *Our Mr. Sun*.
Photo courtesy Wesleyan Cinema Archives.

Frank Capra called the professor, and invited him to be "Dr. Research." Neither man knew it, but Capra was offering him a kind of celluloid/video immortality. Several generations were going to be affected by Baxter's decision, because without him, the Bell Series would not have been half as effective. Ironically, Baxter initially said *no.* "I have no idea how to teach science," he protested, "and I'd probably look stiff on camera."

Capra did his best to persuade Baxter otherwise. "Be yourself," he advised, and bring the same kind of enthusiasm you brought to your Shakespeare shows." Baxter finally agreed, and Capra's instincts proved sound. Apart from his engaging personality, Baxter had a phenomenal, perhaps even "photographic" memory. He could rattle off complicated scientific terms as if he had studied them his whole life. And though the subterfuge was unintentional, Baxter's doctorate in English helped foster the illusion that he really was a scientist. Viewers young and old automatically assumed the "Dr." in front of his name came from a scientific discipline.

The next important actor to be cast was "Mr. Fiction Writer." Why he was called "fiction writer" as opposed to just "A writer" is unknown. An educated guess might be that fiction is largely a product of the imagination—and Fiction Writer was going to be the creator and "master" of the second, "Magic" screen. Fiction Writer was also supposed to be a stand-in for "we the people," that is, the average American. He'd ask questions and make comments that Capra assumed would be on the minds of the viewers.

Actor Eddie Albert was cast as Mr. Fiction Writer. Although he's best known as "Oliver Wendel Douglas" in the '60s rural sitcom *Green Acres*, he actually carved out a good career as a movie actor, mainly in supporting roles. He too had an upbeat, engaging personality, and there was real chemistry between him and Frank Baxter. In fact, Lydia Morris Baxter, the professor's daughter, notes that her father liked Albert very much. There's a mutual regard that can be seen on screen.

Eddie Albert was born Edward Albert Heimburger in 1906. After graduation from college, Albert started a business career, but the Great Depression of 1929 altered his life forever. He tried many different professions, including insurance salesman and circus trapeze performer. A man of many talents, he even had a stint as a nightclub singer.

Albert drifted into radio after he moved to New York in 1933. He co-hosted the radio show *The Honeymooners—Grace and Eddie Show*, which ran for three years. He changed his last name to "Albert" for painfully obvious reasons. Too many announcers were introducing him as "Eddie *hamburger*."

It's interesting to note that Albert was one of the early television pioneers. In fact, he wrote and performed in the first teleplay, *The Love Nest*, written specifically for television. It was telecast live from Studio 3H, the GE Building, in New York's Rockefeller Center in1936. There were very few television sets at the time because America was still in the grip of the Great Depression, and sets were so expensive only the rich could really afford them. The cheapest version of a twelve-inch set cost around $445 in 1938. That's roughly $6,000 in today's money.

It took a lot of stamina to be a television performer in the 1930s. The makeup was very heavy and deliberately garish. Those early cameras had problems transmitting certain color values and line contrasts, especially the color white. To look normal on screen the performer had to look in real life something like the Wicked Witch in the 1939 version of the *Wizard of Oz*. The basic makeup was distinctly green, and black lipstick was also heavily applied.

It took so long to put on this "Halloween" makeup, Albert did the job himself. Albert and his fellow pioneers also had to endure the searing heat of early television studio lights. The multitalented actor also appeared on Broadway, and in the late 1930s landed a movie contract with Warner Brothers. His first major feature film role was as "Bing Edwards" in *Brother Rat* (1938), with Ronald Reagan and Jane Wyman. It was a part he had previously essayed on Broadway.

Albert was a genuine hero in World War II, distinguishing himself during the bloody battle of Tarawa in the Pacific Theater. In November 1943, during the height of the battle, Lieutenant Albert, USNR, piloted a landing craft that rescued forty-seven Marines who were stranded and trapped. He later supervised the rescue of thirty more leathernecks. The operation was conducted under very heavy Japanese machine gun fire. In recognition of his service, Albert was awarded the Bronze Star.

After the war, Albert formed his own film company, appropriately titled Eddie Albert Productions. He chiefly produced films for corporations but did lens a very controversial documentary called *Human Beginnings*. It was a sex education film, very frank and almost shocking for the 1940s.

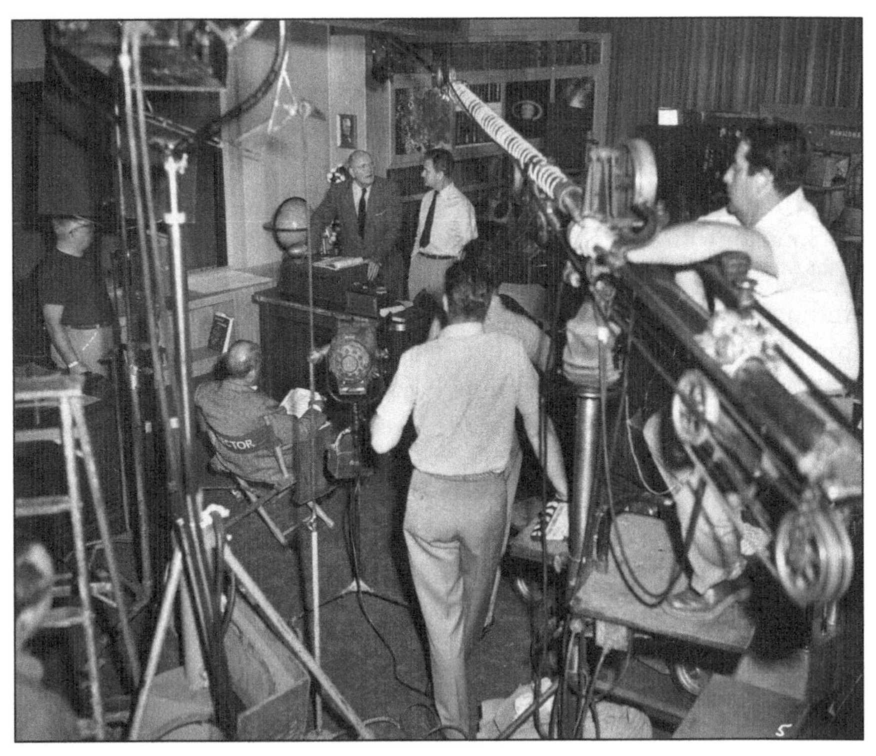

Behind the scenes photo of Frank Capra (in director's chair) filming a scene in *Our Mr. Sun*. Photo Courtesy Weslyan Photo Archives.

50s, he continued his film work. Just before getting involved in, he got an Academy Award nomination for Best Supporting Actor for *Roman Holiday* (1953). The movie is best remembered for being Audrey Hepburn's screen debut. Albert also became a committed environmentalist and activist campaigning against world hunger.

Our Mr. Sun also needed voice actors to bring the cartoon characters to life. "Father Time" was voiced by Lionel Barrymore, one of the most distinguished actors of his time. Barrymore came from a family that was considered "royalty" of the stage. His brother John Barrymore and sister Ethel Barrymore were also famous thespians of stage and screen. Sadly, Barrymore broke his hip twice, and that, combined with severe arthritis, confined him to a wheelchair by about 1938, when he was sixty.

Tough and courageous, Barrymore worked to the end of his life, though the arthritis kept him in daily, and often severe, pain. Lauren Bacall, who worked with him in the classic movie *Key Largo,* remembered how he groaned in agony in his sleep. That he was able to sleep at all was due to studio head Louis B. Meyer, who spent $400 a day to secretly obtain cocaine for the ailing actor. Though often in agony, and sometimes filled with pain-killing drugs, Barrymore always turned in first rate performances.

Our Mr. Sun was Lionel Barrymore's last professional work as an actor. He died on November 15, 1954, a few months after *Sun* was completed. *Our Mr. Sun* would not air until almost two years after the actor's passing.

The role of Mr. Sun himself was voiced by Marvin Miller. Miller, a stocky actor with a rich baritone voice, divided his time between onscreen performances and voice work. His most famous role was that of "Michael Anthony" in the television drama *The Millionaire.* The show, which aired from 1955 to 1960, featured Miller giving $1 million checks to unsuspecting people. The drama then showed how the character's life, and even personality, was altered by the sudden wealth.

The Bell films are probably his best-known voice work, but there was one other. Miller was the voice of "Robby the Robot" in the sci-fi classic *Forbidden Planet*, a movie which has been compared to Shakespeare's *The Tempest.* It was the first literally cerebral thriller, since one of the main characters, Dr. Morbius, unleashes a monster from the Id—his inner subconscious.

Our Mr. Sun was nationally televised by CBS on November 19, 1956. The film aired at 10 p.m., which yet again underscores the fact it was meant for a general, not necessarily child, audience. By any standard the show was a hit, and reached twenty-four million viewers in the United States

Eddie Albert and Dr. Baxter interacting with Mr. Sun and Father Time. Author's collection.

and Canada. It received an audience rating of twenty-five and a thirty-two percent share of the audience. It beat out the competing *Robert Montgomery Show,* but lost out to Lawrence Welk's "champagne" musical program.

It seems like the ad men didn't really know how to market *Our Mr. Sun*. It's plain that both Bell and Ayer, in spite of their ongoing enthusiasm, were still a little nervous about the project. An ad that ran in *TV Guide* provides a good example of those ratings "jitters."

The ad features Frank Baxter and Eddie Albert in profile, looking at each other. Albert is wearing a big grin on his face. A cartoon Mr. Sun is in the center of the ad, looking similar but not exactly like his film alter-ego. Baxter and Albert look as if they are cast members of a new situation comedy. No white-coated "longhair" scientists here! This will be *fun* as well as a learning experience, the ad seems to say.

The positions of the sun, Albert, and Baxter suggest a sitcom of the era. "Don't Miss *Our Mr. Sun*" the ad copy admonishes, "You've never seen anything like it!" In smaller print the ad explains that the show "presents amazing scientific information with the excitement, drama, and humor of

Typical ad pitching *Our Mr. Sun* as entertainment.
Photo courtesy Author's Collection.

popular entertainment. It stars the popular television personalities Eddie Albert and Dr. Frank Baxter."

Now, in the ordinary course of events, the very label "science" would be the ratings kiss of death for any show. So the ad copy underscores the fact that the show "is an exciting combination of dramatic photography and sparkling animation."

The opening credits immediately establish the Capra connection between science and religion by quoting a verse from Psalms: "The Heavens Declare the Glory of God." The soundtrack plays a selection from the fourth movement of Beethoven's Ninth Symphony, Choral, establishing a kind of otherworldly yet triumphant mood. Then we are taken to a kind of studio, where "Dr. Research" and "Fiction Writer" are putting together the final touches of the *Our Mr. Sun* show. This adds an almost surreal aspect to the proceedings because the audience is supposedly "eavesdropping" on a rehearsal of the upcoming show—which of course, it isn't, because this "rehearsal" *is* the show.

Mr. Fiction Writer is seen banging away at his typewriter while Dr. Research reviews the scientific graphics on a small screen. "How was the sun born?" Albert asks mischievously. "We don't know," Baxter replies. Albert suggests that the show would be more effective if they combine "your science and my magic" to the developing show. "We'll open up the curtains of our imagination," Albert declares, while pulling on the chords of real curtains that hide a large "magic" screen.

As the plot unfolds, Mr. Sun and Father Time appear on the "magic screen." Mr. Sun has a low boiling point, and soon is driven to white-hot anger when he recalls how modern man treats him. He nostalgical-

Mr. Sun and Father Time. Photo courtesy author's collection.

ly remembers when various ancient cultures thought he was a god. He was Shamash, Mithras, Ra, and the Greek deity Apollo. But one of those Greeks, Anaxagoras, said that the sun was a hot stone and not a god.

This began man's quest towards logic and reasoning, which in turn eventually produced the modern scientific age. Now the sun feels like some specimen, but Baxter and Albert try to convince old sol that all this research is beneficial. The sun is the "head of our family" giving us light, heat, and energy, and in trying to understand him we can also appreciate him.

Much of the rest of the program details what we know about the sun, and some of the mysteries that still enshroud "him." Baxter presents his facts with a combination of stock footage, colorful graphs, charts, and animation. Occasionally, film clips of distinguished scientists are shown; even Dr. Menzel, the thorn in Capra's side, makes a cameo appearance.

More complicated subjects are further explained by the use of clever "fact" animation. Thermo the Magician, pointed mustachios bristling, describes the carbon cycle, whereby hydrogen atoms are converted into helium atoms by fusion with carbon atoms. In the process energy is created. This happens at 30 million degrees Fahrenheit—"this is not children playing with matches" says Thermo sagely.

The first half or more of the film holds up scientifically even today. The last twenty minutes or so are the most dated. Because of a growing world population, finding enough food for all is a great problem. *Our Mr. Sun* speaks excitedly of chlorella, algae that can "produce about ten times the edible material, per unit area, as a typical crop." Alas, a few years after *Mr. Sun* aired, chlorella was found to be a dead end, at least in terms of feeding the world's hungry. It simply cost too much to produce commercially.

Solar cell technology was in its infancy in the 1950s, but *Our Mr. Sun* does a pretty good job in explaining the basic concepts. Baxter and Albert are so positive, so upbeat, about man's solar energy future, that you find yourself caught up in the enthusiasm, even though you know most of what they predicted did not come to pass.

Capra made sure his religious message was well embedded into the film's narrative. Wise old Father Time tells the audience, "somebody (i.e. God) loves you" because the earth is not too cold and not too hot, at least when it's compared to other planets in our solar system. And Father Time later remarks that "It's right that you should want to know, or the good Lord would not have given you that driving curiosity."

It's interesting to note that critics and public alike gave little notice to the religious "chestnuts" that Capra threw in their path. In the 1950s, organized religion experienced an upsurge, probably because it seemed an antidote to the "Godless" atheistic Communism that seemed poised to engulf the world. The Eisenhower era was an age of anxiety as well as an age of prosperity. There are no atheists in a soldier's foxhole, and Americans faced the constant fear of nuclear annihilation. Even secular films like George Pal's sci-fi classic *War of the Worlds* ended with a grateful humanity singing hymns to God for their deliverance from Alien conquest.

Some reviews actually *liked* the religious references. The *San Francisco Chronicle* enthused that *Mr. Sun* "was the hottest thing on television... best of all a few spiritual overtones: 'Measure the outside with mathematics, but measure the inside with prayer. Prayer is research too.'" The quote comes from Father Time, and when he's saying it, the visuals display a cross for added emphasis. One can well imagine Menzel's reaction to such theology.

One wonders what Frank Baxter really thought about Capra's religious obsession. Certainly, he mouths the holy platitudes with real sincerity and conviction—but then again, Baxter was a good actor as well as a teacher. Today, his daughter Lydia Morris says that the professor wasn't very religious. "He was raised a Baptist, but attended church services to please his mother, After my parents got married my mother, who was a Quaker, wanted him to join the Church of England when they were abroad, but my father said no, church meant nothing to him." When asked specifically about her father's religious beliefs, Lydia states firmly "he was an agnostic."

Capra was gratified by the praise, and *Our Mr. Sun's* great success eased the growing tension between the director and his "bosses," Ayer and AT&T. The corporate giant, seconded by Ayer, was trying to run the Bell Science series like an assembly line, with speed and the "bottom line" of paramount importance. But Capra was doing something new, and the animation sequences simply could not be rushed.

Work soon began on the second Bell Science feature, *Hemo the Magnificent.* Capra was pleased with Bill Hurtz's work, and would probably have given U.P.A. the nod for *Hemo* had the animator still been with that company. But Hurtz had recently switched over to Shamus Culhane Productions, so they got the coveted assignment. Shamus Culhane was a great animator in his own right. Born James Culhane in 1908, he didn't like his first name, so he changed it to the Irish equivalent, Shamus.

Culhane had a long and distinguished career that lasted sixty years. He worked for many important studios, but perhaps is best remembered for his participation in Disney's *Snow White and the Seven Dwarfs*, the first feature-length animated film. Culhane and his team worked for six months on one of the most celebrated scenes in the film, the dwarfs coming home from the mine lustily singing *Heigh-ho* as they trudge along.

Hemo is considered by many as the best of the four Capra Bell films. This time the "magic screen" is populated by some Disney-like forest animals who live in and around some ancient Greek temple ruins. Their "god" is Hemo, a very stylized deity much like Apollo, except that his transparent chest holds a large heart that looks like a valentine. Hemo, like Mr. Sun before him, is portrayed as a skeptic who ridicules man and his efforts to understand the cardio-vascular system. Gradually, though, Dr. Research (a.k.a. Frank Baxter) wins him over with a dazzling array of facts and colorful visuals.

In his autobiography, *Talking Animals and other People,* Culhane freely admits that "*Hemo* was a very difficult picture in that the production was full of crowd shots. Hemo trailed a large group of animals after him, and no matter how skillfully Hurtz composed the film, these animal shots could not be avoided. In spite of the many technical problems the picture was finished on schedule."

Bill Hurtz was the Animation Director on *Hemo*. The project was so big, his staff ballooned to forty people. When the space ran out, Shamus Culhane Productions relocated to a two-story building with an ample parking lot. Capra's name was magic when it came to attracting top talent for the Bell films. Tony Rivera came aboard as storyboard artist, a man who was considered the best in the business. Ben Washam also was hired, a star animator who had worked with the legendary Chuck Jones ("Bugs Bunny") at Warner Brothers.

As always, casting was of primary importance. Dr. Baxter was signed to reprise his role as the all-knowing yet avuncular "Dr. Research." Actor Richard Carlson replaced Eddie Albert as "Mr. Fiction Writer." A handsome man who vaguely resembled the more famous actor William Holden, Carlson was a multitalented individual who was also a competent director and screenwriter.

After returning from World War II service, Carlson found it hard to get decent acting work. His career was revived when he costarred in the filmed-on-location classic adventure *King Solomon's Mines* (1950). But fewer feature movies were being made, partly due to the onslaught of television. Nevertheless, Carlson found his niche in a series of sci-fi and horror films, most notably *The Creature from the Black Lagoon* (1954). Carlson was going to remain as "Fiction Writer" for the remainder of the Capra Bell series.

Sterling Holloway played "Jim," the fumbling if friendly film technician who appears one or two times in the course of the program. Holloway had a very distinctive, high-pitched, and slightly raspy voice, which put him in great demand as a voice actor. His most memorable role is probably the original voice of Disney's *Winnie the Pooh,* but he did scores of productions. As an actor, his unruly shock of red hair made him a natural for comedies. His role in *Hemo* is essentially twofold: to advance the story in the beginning, and to provide a minute or two of comic relief.

Voice actors were especially needed on *Hemo*, because the Magic Screen is crowded with a whole menagerie of animals. Marvin Miller essayed Hemo himself, the deep baritone cadences of his voice underscoring the god's authority and power. The squirrel, turtle, alligator, and rabbit were done by famed voice actor Mel Blanc. Blanc was a master of voices, but probably the most famous are from the stable of Warner Brothers cartoon characters, particularly Bugs Bunny.

June Foray essayed the deer and bird characters in *Hemo*. Her talent for voice work was such people have called Mel Blanc the "male June Foray." She too has a long and very impressive voice work resumé, but is probably best remembered for "Rocket J. Squirrel" of the *Bullwinkle Show*, and the title character in *Crusader Rabbit.*

Now in her mid-nineties, Foray is the last principal cast member of the Capra-Bell Science films who is still alive. She was not on the *Hemo* set itself, but she did her deer and bird impressions in a recording studio. Because of this, she cannot provide anecdotes on Baxter or Carlson. Foray does have warm memories of Frank Capra, who she recalls was a pleasant man and very nice to work with.

Shamus Culhane also recalled Capra in a positive light, and provided a glimpse of Capra at work on the Bell series. The first time the two men met, Capra was sitting in front of a moviola in Culhane's cutting room, looking at some pencil tests of *Hemo*. "I was struck," Culhane wrote, "by the fact that Capra had a pair of eyes as black and piercing as Picasso's, but he had a very relaxed and informal manner."

The director seemed easygoing, nodding quietly when some script change was presented to him. And yet, underneath the surface, there was a blazing intensity to the man, and Culhane suspected he had an explosive

Sicilian temper when he felt things were not going right. Culhane adds that he never displayed his temper on *Hemo*, because the whole team worked hard and things went well.

Magic Screen has its moments, but it's Dr. Research's "fact screen" that really steals the show. Take, for example, Professor Anatomy, who is a rolled-up anatomical chart that springs to life. The heart, essentially a big muscle, is shown to possess four chambers. The Professor shows there are the two "reception rooms" (atria) where the blood enters the organ. It then passes through "little doors" (valves) into the two "living rooms" (ventricles), where it's pumped out. The pumping mechanism of the heart is represented by little "muscle men" who act in concert with one another. The basics are thus shown in an entertaining matter than all can understand.

The animated fantasy is backed up with fascinating reality. When speaking of the capillaries, an animated segment shows that they are so small, individual red blood cells can only pass through one at a time. The actual flow is shown controlled by little "muscle men" who act as gatekeepers on valves. Once the basic premise is established, the film shows microscopic shots of the actual capillaries of live bats, frogs, and hamsters. The animated sequences complement the real-life footage, so viewers can really appreciate what is going on when they are exposed to the "scientific" portions.

There's one scene that everybody remembers in *Hemo the Magnificent*: The confrontation between "Dr. Research" and Hemo himself. After the first half hour explains how blood circulates, the god can stand no more. Hemo is tired about hearing the details of what he calls "plumbing." He demands that Baxter describe the "poetry, mystery, and true meaning of blood" in just two words. Baxter confidently says "sea water," and is proven right. Scenes like that stick in the memory even after the passage of fifty years.

A lot of scientific facts, ordinarily boring, are presented in a lively and entertaining way. Using vivid animation, Dr. Research says the heart can pump a quart of blood in ten seconds. But in ten minutes it could fill "the gas tank of your car," in ten hours fill "a gas truck," in ten days "the average home swimming pool," and in ten years "two ocean-going oil tankers." As Baxter relates the facts, a little cartoon man gasps in awe as he sees a car, truck, a swimming pool, and two ships fill with red liquid.

Once again, Capra can't resist linking science and religion. To him, God is the prime mover behind the wonders and intricacies of the cardiovascular system. The opening credits once again quote the Bible, this time from Leviticus: "For the Life of the Flesh is in the Blood." Beethoven's Third Symphony, "Eroica," is also on hand to give the opening a proper triumphant and exhilarating air.

Capra believed in God, and he certainly was a devout Catholic, but the director was no creationist. *Hemo* talks about evolution as a fact, not theory, though once again Capra weds God and science without the blink of an eye. Millions of years ago the first sea creatures began to crawl out on the land and slowly evolved, from amphibians to reptiles to mammals, until finally man appeared. These creatures changed and evolved for a reason "known only to God."

Now, Capra is skating on thin ice, and he knows it. Why, does this mean that humans were not separately created by an all-loving deity? Sounds suspiciously like one of those "atheistic long haired" scientists! So, he backpedals a bit. Carlson, representing the average person, bristles at the thought. "Now wait a minute, doc," he declares indignantly, "are you trying to say I'm descended from some kind of sea gnat?"

Baxter, all smiles, reassures Carlson : "you have a human spirit that entirely separates you from the animal world." Mr. Fiction Writer is visibly relieved, and Baxter goes on in a more philosophical tone. "There is great mystery and great wonder in the fact that our body—this temple of the spirit—is built of billions of highly specialized individual cells like minute tropical sea animals."

The film concludes with Hemo emphasizing the fact that there is no conflict between science and religion. "One of your greatest physicists, Max Planck," Hemo relates, "said that over the temple of science should be written the words 'Ye Must Have Faith.' Your great Apostle Paul wrote to his new church in Thessalonica: 'Prove all things: Hold fast that which is good.' A scientist says 'have faith.' A saint says 'prove all things.' Together they spell 'hope.'"

When it was broadcast by CBS on March 20, 1957, *Hemo* was a ratings winner, beating out the NBC *Kraft Theater* and the ABC situation comedy *The Adventures of Ozzie and Harriet*. Capra himself began to get stacks of favorable fan mail. As might be expected, many clergy were also happy and told him so.

Hemo the Magnificent was both a critical and popular success. It got rave reviews, and won a Howard W. Blakeslee Award given by the American Heart Association. The program also received the Best Cinematography for Television award at the 1958 Emmys. Capra and company seemed

Baxter meets Dostoevsky and puppeteer Bil Baird on the set.
Photo courtesy Wesleyan Cinema Archives.

on a roll, but nothing lasts forever, especially phenomenal success.

The Strange Case of the Cosmic Rays was the next entry in the Bell Science series. In a sense this was familiar territory for Capra because he had written a treatment on the subject many years earlier for Cal Tech. But though the basic format remained the same, Capra decided to abandon, at least temporarily, the all-cartoon nature of "Magic Screen." He hired famed master puppeteer Bil Baird to create some characters instead.

The original idea was to have a "Sherlock Holmes" try to solve the mystery of cosmic rays, but the character was soon dropped in favor of an entirely different scenario. There would be a contest for the greatest mystery story ever told. Marionettes Charles Dickens, Edgar Allan Poe, and Fyodor Dostoevsky would judge the entries and bestow the coveted "Edgar" award. When Baxter and Carlson arrive at the scene, they try to convince the judging panel that the search for the identity of cosmic rays is the most exciting detective story of all.

Cosmic Rays is the weakest of the Capra Bell series, mostly because the subject itself is too abstract, even esoteric, for the average person. As Culhane admits, "the information on cosmic rays was almost entirely abstract, so Capra fell back on a dizzy assortment of characters." In one scene, a cartoon shows the "First Electroscope Bank of Sciencetown," where electrons are deposited. But these electrons are being stolen by visible ray "bandits."

There's a slinky Mae West thief who personifies "ultra violet" rays, and her "cousin" gamma rays. Eventually they are apprehended. Then there's the "Uranium gang" guys, numbers 234, 235, and 238. A voiceover from Carlson says these are the "Atom bomb" boys. They too are thrown "in jail." But there's still a thief around stealing electrons: a mysterious phantom who Baxter and Carlson call Fagin in honor of Charles Dickens. "Fagin" is invisible, and much of the rest of the program is devoted to finding out his true identity. It ends up "he" is the elusive cosmic rays.

Cosmic Rays aired on October 25, 1957. Try as they might, Baxter and Carlson—through the Capra script—partly fail in making the science intelligible to the average person. The film does have some interesting moments, and the Bil Baird marionettes are fun to watch, but it becomes tedious and confusing when Carlson assembles a crime "dossier" on Fagin/Cosmic Rays that lists a bewildering array of protons, neutrons, Mu

Rehearsal on the set of *The Strange Case of the Cosmic Rays*. Left to right: Dr. Baxter, Richard Carlson, and famed voice actor Mel Blanc. Photo courtesy Wesleyan Cinema Archives.

Mesons, Pi Mesons etc. Only with repeated viewings does at least some of the information start to register. And repeated viewings were one luxury the television audience and the later classroom students did *not* have.

It's a forlorn hope, but *Cosmic Rays* uses music and sound effects to enhance the viewer's experience. Classical music is very much on hand, just as it was in the two previous Bell Science entries. Beethoven's Ninth Symphony, Fourth Movement (Choral) serves well, just like it did in *Our Mr. Sun.* But for *Cosmic Rays,* it's the music of Richard Wagner that really takes center stage. When cosmic rays are bombarding the earth from all directions, the Ride of the Valkyries from Act III of Wagner's *Die Walkure* provides the stirring background music.

But perhaps the most interesting aspect of the soundtrack is the use of the theremin—an instrument that produced "spooky" or "other-worldly" sounds. The theremin was a favorite with science fiction movie producers, so audiences were conditioned to hear this music when something strange or alien appeared on the screen, like a monster or flying saucer. In *Cosmic Rays* the theremin and another instrument called the vibraphone are often used when displaying stars, galaxies, or other outer space phenomena. Carlson's presence heightens the mood, since he was the star of such sci-fi classics as *It Came from Outer Space* (1953).

Perhaps because the subject is so far removed from the lives of ordinary people, Capra takes particular pains to sell *Cosmic Rays* as exciting, fun, and entertaining. The marionette Fyodor Dostoevsky represents the audience. While Edgar Allen Poe and Charles Dickens have an open mind, and listen with interest, Dostoevsky doesn't even like talking about cosmic rays. "Nyet!" he cries, dismissing the whole subject. "Science for me is a big *bore.*"

Dostoevsky as the audience surrogate resists all attempts to win him over in the film's early stages. When Dr. Baxter explains the workings of the electroscope—a simple instrument that detects radiation—Dostoyevsky is unimpressed. "Vat is this, a schoolroom?" he says in a comic Russian accent. "Where is the mystery?"

Gradually old Fyodor is won over. As the story unfolds, he begins to make bets with Poe and Dickens over the "mystery" of cosmic rays, advancing his own opinion about their identity. The wagering shows his resolve is weakening; almost in spite of himself, he is finding science fun. And by the time the film ends, he is completely won over. There's a vote on who should get the "Edgar" award, and Dostoevsky enthusiastically votes yes.

Baxter consults with Carlson in a scene from *The Strange Case of the Cosmic Rays.* Photo courtesy Wesleyan Cinema Archives.

Carlson "Mr. Fiction Writer" is amazed. He expected the Russian to vote "Nyet." But it seems Dostoevsky is a changed man—er, marionette. The identity of the mysterious cosmic rays—which the film calls "Fagin," after the Dickens villain in *Oliver Twist*—is revealed to be the "naked atom." In trying to find the identity of the mysterious cosmic rays, "Fiction Writer" says scientists discovered "material proof" that everything in the universe is essentially made up of the same matter.

Then, Mr. Fiction Writer starts to wax poetic. A beautiful young woman is shown onscreen with long, shoulder-length hair. "The carbon in this young lady's shining red hair," Carlson asserts, "might once upon a time have been stardust from an exploded supernova, or from the Southern Cross!" As he's speaking, Beethoven's Ninth comes on the soundtrack with the appropriate "extraterrestrial" chorus.

Impressive, but Dostoevsky brings up a valid point. The cosmic ray story "For highbrows this would be exciting—maybe. But what is the message for the common people who work, love, and hate?" This is a set up for Dr. Baxter to deliver the knockout blow to all skeptics who think that science is not relevant to the average person. "These cosmic bullets (rays)," explains Baxter, "could be partly responsible for evolution."

This is the second time that evolution was mentioned in the Bell Science series: the first was in *Hemo the Magnificent*. It is obvious that Capra is no creationist, nor does he think the earth is only 6,000 years old. He believes in evolution, but there's no way of knowing how he reconciled the literal Bible account with Darwin. Perhaps he was more of a deist in this instance, believing in a "watchmaker" God who set the laws of nature in motion, then let them alone to evolve according to plan.

Dostoevsky is now a firm believer in science, and (Capra hopes) so is the audience. The Russian votes in favor of awarding the cosmic rays story an Edgar. But Charles Dickens is not so eager as his volatile colleague. He says a "vast ocean of truth still lies before you undiscovered. We should invite these gentlemen back in fifty years to tell us what new chapters they've added to their cosmic tale."

"And who will add these new chapters?" Carlson asks, but answers his own question: "The young." he exclaims. "The inheritors of all the knowledge and culture of the ages, all the dreams and hopes of those that came before them. What an exciting time to be young!" Dr. Baxter chimes in, making a last pitch for a nuclear future. "What a glorious opportunity to add to man's history, to harness nuclear fires for the benefit of man."

Capra can't resist a final plug for the Almighty. In the final moments, Dr. Baxter says when we learn more about the "creation," we get closer to the "Creator." The three author marionettes toast "stardust," and a picture of a galaxy fills the screen, while once again the stirring strains of Beethoven's Ninth Symphony fills the ear. The overall effect is uplifting and positive. Man *will* solve the riddles of the universe, and he *will* master the atom for peaceful uses.

Like all of the Bell Science series, *Cosmic Rays* is optimistic about how science can change our future for the better. That's part of its charm, and one wishes we had that kind of positive attitude today. Nevertheless, parts of the Bell Science films are naïve about how humans can manipulate nature for its own benefit. When Charles Dickens says a mystery story must have scope, Richard Carlson practically bubbles over with enthusiasm. "*Scope* he says!" Carlson declares. "The cosmic rays case has revolutionized the modern world!!"

Mr. Fiction Writer then "proves" his case by giving us all a glimpse of our wonderfully radioactive nuclear future. While a montage of images flashes on the screen, Carlson's voice-over lists "future nuclear plants, nuclear (powered) subs, nuclear planes, radioactive drug store, and isotopes for medicine..." There's a pause, and Carlson says "Hold on to your hat!" The screen then fills with frightening images of atomic bomb detonations, complete with the all too familiar mushroom clouds. "How's *that* for scope?" Carlson crows triumphantly.

Baxter speaks of the three forces of the universe: gravitational force, electromagnetic force, and nuclear force. But it's nuclear force that merits Baxter's full attention. We have to look in the heart of the atom, in the nucleus, to find this nuclear force. It's the "awesome force that man is now unlocking for his own use." Nuclear force is one of the most powerful elements in the universe, and it holds the protons and neutrons locked in the core of the nucleus.

This nuclear force will be tamed for man's peaceful use. Dr. Baxter proclaims that nuclear power is the "force that will surely change the way of life for every man, woman, and child on earth." Baxter's original audience in 1957 knew all too well the risks and dangers of nuclear war. But Capra and Bell were offering an alternative to mass destruction. This is very much the message in the 1950s Disney cartoon documentary *Our Friend the Atom.*

In *Cosmic Rays,* humans have mastered the secrets of the atom, and are aware of the dangers of radiation. Disasters like Three Mile Island and Chernobyl are unthinkable to Capra and company. The director may have been at odds with the scientific world, at least when it came to matters of religion, but he shares their view of the limitless and problem-free possibilities of atomic power.

Though Capra still got fan mail, and many critics praised *Cosmic Rays,* negative reviews were on the rise. The director soldiered on with the last Capra-Bell Science offering *Meteora: The Unchained Goddess. Meteora* was a very effective documentary on the weather, and much of it holds up even today. The weather goddess must be a woman, Capra decided, because "weather is so unpredictable, at times so beautiful, and at other times so terrifying." No producer in his right mind would make public statements like this today, but this was the fifties.

Actually, apart from the religious angle, all of Capra's Bell films are very much products of their time, especially when it comes to women. As previously mentioned, in *Our Mr. Sun* the sun shines down enough energy to power a lady's sewing machine, washing machine, etc. The assumption is, of course, that most women are housewives, since that is their "natural" role in life.

In *Meteora,* the goddess starts to fall in love with good old Dr. Frank because he's the only one that seems to understand her. Finally, she plucks up enough courage to ask him to marry her. With typical 1950s sensibilities, she reasons that though a woman couldn't ask a man to marry her, a goddess could. When he says "I'm already married," she is momentarily crushed. But the production ends on an upbeat note, as Carlson recalls "Dr. Research" had predicted fair weather. Baxter and Carlson look out a window where it's pouring rain. Meteora and her attendant deities, happy that modern science finally stumbled, break into raucous laughter.

The film opens with a swirling mass time-lapse clouds, while a female voice—Meteora herself—says that some mortals are trying to understand the weather. Laughing, the goddess says "Let's watch them make fools of themselves." After a bit of preliminary banter, Dr. Research and Fiction Writer turn to the Magic Screen and there they meet Meteora and her attendant deities.

Baxter first explains wind and how it's influenced by the rotation of the earth. To further explain matters, Dr. Baxter turns to a cartoon colleague: Professor Coriolis. The name was taken from the real-life Gustav de Coriolis, a nineteenth century engineer who identified the force that acts on moving objects due to the earth's rotation.

The cartoon Coriolis, with academic mortarboard on his head and a comic French accent, demonstrates why winds curve, and does so by using a live action carousel and two boys throwing a ball. It's a brilliantly conceived segment, and some science teachers show it in classrooms to this day. The Coriolis segment shows Capra at his best, explaining scientific concepts in an entertaining and thoroughly understandable way. Here, as elsewhere, he really is building bridges between the scientific community and the general public.

Meteora, beautiful but petulant, is not convinced these "mortals" can stand up to her attendant deities. Each god is a personification of an element of weather. There's Borealis, god of the winds, who gets insulted when Carlson irreverently calls him "windy." There's also a cloud god, and one for rain, snow, and so on. Baxter and Carlson take them on one by one, reducing them to myth by showing how weather really operates in nature

Aided by a combination of cartoons and real-life visuals, "Dr. Research" Baxter explains how clouds are formed from dust and water, and how positive and negative electrical charges form lightning. He also talks about the concept of weather "fronts," and the drawing of isobaric maps.

Richard Carlson points out a huge tornado in *Meteora: The Unchained Goddess.*
Photo Courtesy Wesleyan Cinema Archives.

There are some interesting, even exciting, moments when actual footage of hurricanes and tornadoes is shown. At one point, a B-29 weather data plane is seen battling through a hurricane to reach its calm "eye."

The film is dated towards the end, when a clip is shown of the Weather Analysis Center in Suitland, Maryland. There, a team of "weather psychiatrists" are making great strides by using new tools like computers. But what was state of the art in 1958 is positively prehistoric in the twenty-first century. *Meteora* proudly showcases a computer world with vacuum tubes, reels of magnetic tape and dot-matrix printouts. The computers shown in the film could probably do 2,000 calculations a second and had twelve K of memory. Today, an average computer does about 600,000 calculations a minute and has three gigs of memory.

Meteora is also understandably dated on data gathering. Meteorologists are seen sending instruments aloft in weather balloons. Today, of course, we have weather satellites. In 1958, the year *Meteora* was made, the United States launched its first satellite, Explorer I. The "ancestor" of today's sophisticated weather satellites was Tiros I, which was launched by NASA in 1960.

In the 1950s, both scientists and the general public were somewhat naïve about man's ability to control the forces of nature, and the film underscores this fact. Hurricanes are always a problem, and *Meteora* does a good job in explaining how they form. But to control these terrible hurricanes, the movie naively suggests such ecological disasters as pouring oil on the surface of the ocean and lighting it. No thought is given to the pollution of the oceans, or the terrible harm that burning oil would do to marine mammals, birds, and sea life.

Capra and company are much more prescient about the subject of global warming. "Even now," Baxter says, "man may be unwittingly changing the world's climate through the waste products of his civilization. Due to our release through factories and automobiles every year of more than six billion tons of carbon dioxide, which helps air absorb heat from the sun, our atmosphere seems to be getting warmer."

Carlson, looking puzzled, asks "Is this bad?" Baxter replies:

> *Well, it's been calculated a few degrees rise in the Earth's temperature would melt the polar ice caps. And if this happens, an inland sea would fill a good portion of the Mississippi valley. Tourists in glass bottom boats would be viewing the drowned towers of Miami through 150 feet of tropical water. For in weather, we're not only dealing with forces of a far greater variety than even the atomic physicist encounters, but with life itself."*

Meteora: The Unchained Goddess was Capra's last Bell Science film. Ayer wanted more control, and he was unhappy that each film cost $400,000—well over the original budget. They only saw the "bottom line," and had little appreciation of the complexities of making such movies. Capra had long chafed over Ayer's repeated attempts to shackle his creative process.

Shamus Culhane was also initially unhappy, especially when he realized his production company lost a great deal of money on *Hemo* and *Cosmic Rays*. On *Hemo*, for example, Shamus Culhane Productions received a flat rate for the animation. But Capra and Hurtz also created a bonus system. Culhane's animators would get additional money for footage with three characters, a high bonus for a four character scene, and so on.

It wasn't Capra's fault, but this system differed from the day-to-day analysis of the budget and finances that Culhane used when producing television ad spots. In a project the size of the Bell films, you needed close scrutiny of the finances, but the bonus system didn't allow for that. Many months after *Hemo* finished, Culhane discovered his company had sunk themselves into a financial pit by doing the show.

Culhane decided not to confront Capra on this because even if the director was sympathetic, and he probably would be, any extra money would come from a tightfisted Ayer. Besides, in the advertising world of Madison Avenue, seeking additional funds to cover losses after a

"Hail Baxter, Emperor of Education!" Dr. Baxter in costume in *Meteora The Unchained Goddess* Photo Courtesy Wesleyan Cinema Archives.

project is more than bad form—it's a cardinal sin. So Shamus Culhane swallowed hard, bit his tongue, and tried to put the incident behind him.

Things improved for Culhane on *Meteora.* As he put it, "Without the crowds of animals we had to draw for *Hemo*, and the intricate designs of abstract animation which made *Cosmic Rays* so difficult, it looked as if we could make a good profit for the first time." Unfortunately, though the work was easier, Culhane lost money on this one too.

For a time, Capra was bitter because he felt he had been all but booted out by an ungrateful Ayer and, perhaps by implication, Bell Telephone itself. Certainly, as a big-name Hollywood director, he was not used to working with an advertising agency constantly looking over his shoulder. He would call Don Jones, the Ayer rep, and casually mention the need for another day's shooting, or a change in the script. The request in hand, Jones would go back to Ayer and ask for $10,000 or $20,000.

As Shamus Culhane described it later: "He might as well have asked for a pound of bone marrow from each chairman of the board. While Jones managed to get the money, there was grumbling on both sides." *Meteora: The Unchained Goddess* is a fine film, one of the best of the series, but it suffered in comparison to the *Hemo* "gold standard." Critics seemed to be bored by the "magic screen" format, and there were more negative reviews.

Capra wrote privately that "Everybody knew everything but the man who was supposed to make and deliver the films at fixed cost. I looked in the mirror and saw a pigeon." All true, perhaps, but there were other factors involved. Simply put, Capra wanted to get back into feature films, and his agent agreed.

In time, when some of the bitterness left him, Capra looked back with pride on his contribution to the Bell Science canon:

> *By weaving together live scenes, fantasy, traceries of diagrams, animated cartoon character, puppets and—above all—humorous illustrative parables, metaphors, similes, and analogies, we reduced the complex to the simple, the eternal to the everyday. In short, though it took five years, I built a small bridge that spanned the gulf twixt scientist and commoner.*

7
Bell Science: The Warner Years, and a Disney Postscript

Although he probably was thinking about it for quite a long time, Capra's departure from the Bell Science series seemed sudden to many. Shamus Culhane was one of those who felt left in the lurch. His company was in the final stages of *Meteora: The Unchained Goddess* when Capra announced he was leaving the series. As far as Culhane was concerned, the director was cutting the rug out from under them. Dumped, just like that, even after they had sustained financial losses.

Culhane was understandably upset. "What the hell was I going to do with forty or fifty employees, to say nothing of my parking lot?" he asked. But there was still a hope that Shamus Culhane Productions would take over from Capra and continue to make Bell Science films. Why not? Culhane and his crew had three years of experience in making highly specialized animation. They had even created the best of the series: *Hemo the Magnificent.*

Shamus Culhane rushed to Philadelphia and made an appearance at the offices of N. W. Ayer & Son. Culhane sensed there was something wrong because he was not allowed to see Don Jones, the Ayer man who had been with the series since its inception. Instead, he was ushered into the office of some Ayer flunky who knew nothing about the Bell films or Culhane's part in them. Culhane recalled him as a "pompous pipsqueak" who flatly refused to even consider giving the production contract to his studio.

Culhane was told that a "large studio" with a large animation facility had been given the assignment. The Ayer man was reluctant to divulge which studio it was, but finally admitted the mysterious entity was Warner Brothers. Shamus Culhane was disappointed, but in many respects the decision was the right one. Warner Brothers did have an outstanding animation division, and a stable of outstanding talent. Without doubt the dean of the Warner Brothers animators was Chuck Jones, creator of many memorable characters, perhaps the most famous being Bugs Bunny.

Shamus Culhane mentions in his autobiography *Talking Animals and Other People* that "N.W.Ayer gave the work to Warners, and they turned out a turkey." This is absolutely not true; in fact, the Warners films were among the most imaginative, even innovative, of the whole Bell Science series. But Culhane had a right to be upset. He lost a lot of money on the last Capra entry, *Meteora: The Unchained Goddess.* In fact, he was in such dire financial straits that he was forced to close the Hollywood branch of his production company and retrench in New York.

Warner Brothers was one of the major film studios of the Golden Age of Hollywood, roughly from 1930 to the early 1950s. It was founded by Sam, Jack, Harry, and Albert Warner, who were of European Jewish extraction. Warner's wasn't afraid of innovation; their movie *The Jazz Singer*, released in 1927, is considered the first sound "talking" picture. Even though only the musical segments were in sound, the movie created a sensation and sounded the death knell of the silent picture.

By the late 1920s, Warner Bros. had acquired a studio in Burbank, in California's San Fernando Valley, which still exists. It was the ultimate "dream factory," a 110-acre lot with twenty-nine sound stages and a twenty-acre back lot of standing sets. In the 1930s, Warner was home to some of the greatest stars of the "Golden Era," including Bette Davis, James Cagney, Paul Muni, Humphrey Bogart, Edward G. Robinson, Errol Flynn, and Olivia de Havilland.

Gateways to the Mind (1958) was the first Warner entry in the Bell Science series. It's full title, *Gateways to the Mind: The Story of the Human Senses*, pretty much describes the subject matter. Frank Baxter is again on hand, but this time he's exploring sight, touch, taste, smell, hearing, and much more besides. As Geoff Alexander wrote in his book *Academic Films for the Classroom*, the Warner films "did not overly proselytize, relied less on animated characters interacting with Dr. Baxter, and utilized the set design as almost a character itself, as exemplified by William Kuehl's sound stage set for *Gateways to the Mind....*"

Studio head Jack L. Warner selected Owen Crump as the producer for four of the films, and he was also given the assignment to direct three of them. Crump was a handsome, talented man who sported a Clark Gable-like mustache. He was also a writer and director, and by the early

1950s he was one of the most respected men on the Warner's lot. Certainly Jack Warner, who had a well-deserved reputation as a ruthless business man, seemed to like him.

Owen Crump (1904-1998) got his start as a radio announcer in Louisiana. He travelled to California and started working as a director and writer for FFWB, the Warner Bros. radio station. By 1941 he was working for the Short Subjects Department at the Warner Bros. main lot in Burbank. The most notable phase of a long and distinguished career came during World War II, when he became a leading founder of the First Motion Picture Unit.

One day, early in 1942, the chief of the Army Air Forces (AAF), General Hap Arnold, came to Jack Warner with a problem. "Jack," the general complained, "everybody wants to be a pilot. I can't get anybody interested in being a gunner, or crew chief, any of it. Can't you make some short subjects that would make heroes out of these people so everybody will be important so we can establish the crew positions?"

Warner immediately picked up the phone on his desk and called Crump in the Short Subjects Department. "Owen," Warner informed, "you just joined the Air Force." Crump laughed, and said "I *what?*" Warner repeated, "You just joined the Air Force." Warner's word was law, so it was a done deal. Crump's first production, *Winning Your Wings*, starred actor Jimmy Stewart, himself an Army Air Force Lieutenant at the time. It was enormously successful; more than 150,000 Army Air Force enlistments could be directly traced to the film.

For most of the war, Crump's First Motion Picture Unit, or FMPU, was housed in the old Hal Roach Studios in Culver City. For the most part, the FMPU was staffed by some of the finest actors, writers, producers, directors, cameramen, and behind-the-scenes technicians that Hollywood could offer. Ronald Reagan, then an actor but later the President of the United States, was one of those who worked at the FMPU, which was playfully dubbed "Fort Roach."

In the early 1950s, Crump directed a powerful semi-documentary of the Korean War. His film unit actually went to the front line, filming actual combat as it took place. Crump recruited some real soldiers to carry the story along, including a nineteen-year-old Private First Class (Pfc) Ricardo Carrasco. Though a real soldier, Carrasco was called upon to do some acting. In the film, he "dies" in combat, underscoring the terrible loss of war.

Carrasco did what he was told, but he was unhappy about being a temporary actor. He felt guilty about playing war while his friends were fighting and dying in a real one. He continually begged to be released from the film, asking that his death scene be shot quickly so he could return to the front. Hal Wallis, a Warner producer, saw the young man's star quality and offered him a contract, much like WWII hero Audie Murphy had received in 1945.

But Carrasco refused, and filmed his death scene on the morning of July 6, 1953. As soon as it was completed, the young private went back to his unit. At 11:25 p.m. that same day, real life tragically imitated art. Pfc Carrasco was killed by a mortar round at Pork Chop Hill. The film death scene was reshot, so Carrasco's family would be spared the horror of seeing their son "die" on screen.

This film, which was titled *Cease Fire*, was the only production ever allowed to make a movie in the front lines of a war zone by the Defense Department. Crump returned to Warner Bros., where he continued to write scripts for documentary shorts until he was tapped to do the Bell series.

In *Gateways to the Mind*, as well as the other three Warner/Bell Films, the opening credits proclaim that it is "Produced Under the Personal Supervision of Jack L. Warner." That means he was very interested in the series: so interested he was looking over everyone's shoulder to see that the films would conform to his exacting standards. He only did this when genuinely committed to a project. Another film that was under his "personal supervision" was the Oscar-winning *My Fair Lady*, (1964) with Rex Harrison and Audrey Hepburn.

Jack Warner was an enigma to many in Hollywood. He was a Republican who supported the New Deal programs of Democratic President Franklin Roosevelt. Warner was a fervent anti-Nazi, and made anti-fascist films long before America entered World War II. He was also combative, occasionally vindictive, and constantly at odds with both family and employees.

In the Golden Age, movie stars were usually under seven-year contracts. If they refused to do a role, they could be suspended without pay. Warner's battles with some of his top stars were legendary. They often ended up in court. In 1943, for example, Olivia de Havilland took Warner Bros. to court because the studio had added six months to her contract—the times where she had been on suspension. The case went all the way to the California Supreme Court, and the actress won her suit. Thereafter, Warner's—or any studio—couldn't arbitrarily tack on extra contract time.

Though he wasn't a very likeable man, Warner had an instinct for producing good pictures—pictures that were hits with both critics and audiences alike. Warner also wasn't afraid to at least occasionally risk his money by making a "prestige" picture for the sake of "art." A good example of this is the big budget historical epic *Juarez*, which detailed the story of the famed Mexican president and his struggles against French occupation in the 1860s.

The Bell films were something of a family affair for Jack L.Warner. Jack M. Warner, his son and only child, also played a big part in both *Gateways to the Mind* and the later *Alphabet Conspiracy.* Though he had a different middle name—Milton—the younger Warner was called "Junior" around the lot. Assigning Junior to the Bell films was not nepotism: he had extensive experience in Warner's Short Subject Department. For the Bell shows, his official title was "Executive in Charge of Production," a job he shared with Walter Bien.

It was the Warner Bros. Industrial and Commercial Films Division of the studio that actually made the quartet of Bell films. Before Bell, they were primarily engaged in making television commercials for beer, automobile manufactures, and other consumer products. This was something different for them. In 1993, Jack Warner said that these films were "my special pride and I still look back on it with great pleasure."

There's an interesting story about Warner and his son that relates to the Bell films. In 1958, the year that *Gateways to the Mind* was released, Jack L. Warner was vacationing in Cannes, in the south of France. On August 5, 1958, while returning from an evening of baccarat at a casino, Warner's Alpha Romeo roadster collided with a coal truck. Warner somehow was thrown clear, though his car burst into flames upon impact.

The movie mogul was badly hurt, and for a time it was thought he might succumb to his injuries. When Junior visited Warner's hospital room, the older man seemed happy to see his son. Though weak and barely able to talk, he mentioned to his son how *proud* he was of the Bell Science films. Though he did survive, this was practically a deathbed declaration. Jack L.Warner must have been very impressed with the Warner Bell films.

Gateways to the Mind is very imaginative. It takes place on a sound stage on the Warner lot in Burbank. Giant sculptures of an eye, nose, ear, and mouth are placed strategically around the sound stage set. The rest of the space is taken up with the tools of filmmaking: cameras, lights, sound booms, and various pieces of equipment. As the film opens, grips, cameramen, and other crew walk about, seemingly oblivious to the audience and their "eavesdropping."

The almost surreal "film within a film" aspect of *Gateways* is another intriguing aspect of the production. Early in the film, when the opening credits are still unfolding, we see a makeup artist going over the face of an actor in costume. Later, we see this actor playing the ancient Greek philosopher Aristotle in a staged vignette that has him discussing the senses with his students. The audience also watches as a stagehand carefully assemble a rectangular optical illusion—an illusion that won't appear until about a half an hour into the program.

The sets and actors might change, but Dr. Frank Baxter is on hand to provide an entertaining commentary. He's a familiar, reassuring figure, clad as always in his gray flannel suit. We first see him coming down from a camera boom as his voice-over proclaims, "The sound stage, with its gadgets and equipment, will serve as a kind of workshop for our science story." The various film crewmen challenge Baxter, asking him questions in order to better understand the concepts he is explaining. They also occasionally help him with experiments that demonstrate the senses. They are a kind of collective "Fiction Writer," there to be surrogates for the viewing audience.

The illusion of looking "behind the scenes," film within a film, is underscored when Dr. Baxter is occasionally interrupted by what looks like a director, sitting on a folding chair with what looks like a script on his lap. A camera and cameraman are right by his side, looking into the eyepiece as if filming, or about to film, Baxter and his comments. Yet the cameraman, who is identified as "Hal," is really character actor Karl Svenson.

"The Aristotle set is ready, doctor," the "director" says at one point. Baxter walks over to the set where Aristotle is shown discussing the five senses with his pupils. Actor Peter Brown, a prolific performer, plays "Utimus," one of the ancient Greek students. He was a Warner Bros. contract player at the time, but is best known for a number of western television series, notably "Deputy Johnny McKay" in *Lawman* (1958-1962).

Utimus says that "as meat and wine are nourishment for the body, the senses provide nourishment for the soul…All knowledge must come from the senses." The "master" Aristotle nods his head sagely, and speaks of the five senses—seeing, hearing, touching, tasting, and smelling. This provides the backdrop for Dr. Baxter to add a modern perspective. He declares we have more than five senses, and we don't just smell with our noses, etc.

Once again, the "film within a film" technique comes into play, as we see the actors who had just performed in the Aristotle vignette casually walk off the set and presumably to their dressing rooms. Baxter and a stagehand named "Joe" talk while the actors walk just behind them in plain view.

"Wait a minute, Doc!" ("Hal" the cameraman interrupts.) "After twenty years behind this camera, you're not going to tell me I don't see with my eyes?"

"In the strict sense, you don't, Hal." replies Baxter. "Your eyes, like your other sense organs, simply relay impressions from outside and relay them to the brain."

This exchange sets up an introduction to the animation portion of the film. The animators in *Gateways* are actors, though they seem to display "their" work—really the labor of Warner's animation artists Chuck Jones and Maurice Noble.

Both Jones and Noble worked for Walt Disney, but Jones only had a brief stint in the Mouse kingdom, contributing to *Sleeping Beauty*. Noble did layouts at Warner, which is basically the background environment of a cartoon. Both artists are associated with the classic Warner cartoons of the forties and fifties, which are populated by a whole menagerie of mem-

Dr. Baxter discussing sound and hearing on one of the dramatic sets of *Gateways to the Mind*. Photo: Author's collection.

orable characters: Bugs Bunny, Daffy Duck, Wile E. Coyote, the Road Runner, Elmer Fudd, Porky Pig, and many, many others.

Noble was one of those Disney workers who went out on strike in 1941. "Uncle" Walt was traumatized by the event because he felt his employees were "family." After the strike was settled, and Disney was unionized, Walt made sure he would punish those who had been so (in his eyes) disloyal. Noble was assigned a new "office" that had literally been a broom closet! The layout artist stuck it out, so eventually he was given the boot.

Luckily he didn't stay unemployed for too long. Noble joined the army in World War II, and was attached to the Signal Corps. He was attached to a film unit on the Warner Brothers studio lot. The unit was headed by Theodor "Dr. Seuss" Geisel, and famously produced the *Private Snafu* instructional series, which was considered risqué with its mild profanity and semi-nude women. Noble also became friends with Colonel Frank Capra.

Noble seems to have enjoyed his work on *Gateways to the Mind*, considering it very intellectually stimulating. He worked for several months on the project, which was begun in 1957. Although Capra didn't want to have anything to do with the Warner series, according to some accounts, Noble visited his old friend for consultations.

Chuck Jones also enjoyed his work on *Gateways*. "Joe the commuter" is one of Jones's brilliant animation sequences, this one showing how the senses combine to give a picture of our world. "Joe" is walking to work when he sees some strawberries—his eyes tell him they are indeed strawberries, his fingers feel their texture, but his taste determines the berry is sour. Each sense is represented by a kind of elf-like figure, who is loosely based on the swift messenger god Mercury. They dispatch impressions to the brain via the nervous system.

Gateways, like most of the Bell series, deftly combines live action, documentary footage, and animation to tell its story. These elements, seemingly disparate, actually complement each other. The sequence on hearing begins with "Joe," one of the stagehands, playing "chopsticks" on a piano. Baxter walks over to "George" the sound man, who is seated at a console. As if to underscore the subject matter, a huge sculpture of the human ear looms just behind the two men.

George explains he is sound mixing, and Baxter summarizes. "In other words," Baxter relates, "these sounds are first converted to electrical impulses, they balanced and mixed in this console, and then recorded." It's like the human ear and how we hear, except, as Baxter notes, "George's ear is a lot more complicated than his console." The doctor walks over to the animation department, where a series of cartoons clearly demonstrate the inner workings of the ear.

The second half of the film showcases some real experiments with both perception and how the brain stores memory. In the perception segment, Baxter shows a trapezoid with windows that looks like it is moving back and forth. Baxter assures us it's really turning all the way around. To prove his point, Baxter shows a shot of the trapezoid from above. It is indeed moving all the way around.

The secret, of course, is that it looks like a rectangle, but it isn't. Even if we know the truth, our brains refuse to believe it. "The trapezoid is not rectangular in shape," Baxter explains, "but our brain interprets it based on our past experiences." It's a great illusion, and well presented. When a bar is placed on the trapezoid, the bar seems to go fully around, but the trapezoid itself still goes "back and forth"

Dr. Baxter says that this illusion and many others were created in the Perception Demonstration Center at Princeton University. More optical illusions are shown, including a distorted room that seems to make a little boy taller than a grown man. Dr. Hadley Cantril makes a filmed appearance at this stage, discussing the optical illusions and how they represent "profound psychological principles."

Dr. Cantril's delivery is stilted, and he speaks in a monotone. When Cantril says that the distorted images we've just seen are "amusing illusions," one gets the impression that the psychologist never laughed in his life or found *anything* amusing. His "performance" is excruciating, though mercifully brief. He harkens back to the "bad old days" of early science television, with dull, ill-at-ease academics going through the motions at a blackboard.

Actually, Cantril's career is much more interesting than his brief minute or two on *Gateways to the Mind.* A trained psychologist, he joined the faculty at Princeton in 1936. He became interested in studying the effects of media on the general public. Dr. Cantril wrote a book called *The Invasion of Mars*—an academic study of Orson Welles's 1938 Halloween broadcast of *War of the Worlds.* The radio show is remembered for triggering a real-life panic among many on the east coast, who thought it was a real alien invasion.

Cantril also became very interested in public opinion research. In 1940, just before the outbreak of World War II, he founded the Princeton University Office of Public Opinion Research. The Roosevelt administration found the data the Research Office collected was invaluable, especially when trying to gage the American public's feeling on the war in Europe. Cantril's work also had an effect on the course of the war.

In 1942, Cantril gathered a small scale sample of French Vichy officials in Morocco just prior to "Operation Torch," the Anglo-American invasion of North Africa. Vichy was the French puppet-government affiliated with Nazi Germany, but it was hoped that officials might switch sides. The "poll" discovered that there was a very strong anti-British feeling among those Gallic government types. The information strongly influenced the Allies landing patterns, with the Americans landing near Casablanca to prevent any British-French friction.

Cantril also founded the Institute for International Social Research with Lloyd A. Free. The Institute provided data on public opinion for both the Unites States and the world. Sometimes the U.S government used the information, but sometimes they did not. President Dwight Eisenhower okayed a secret CIA plan to overthrow Cuba's Fidel Castro. The intelligence community seemed to think that the Castro regime was hated and on the verge of collapse.

In 1960, Cantril and Free conducted a poll in Cuba, which plainly showed that most Cubans supported Castro and his revolution. Unfortunately, 1960 was a presidential election year, and Eisenhower was in the last months of his term before retirement. Somehow, Cantril's Cuban poll information got lost in the shuffle. Apparently, the new President, John F. Kennedy, never saw the data. He gave the green light to the projected invasion of a small group of Cuban exiles. The result was the Bay of Pigs, one of the worst foreign policy debacles of the 1960s.

As *Gateways* draws to a close, Baxter showcases two interesting segments. One is on sensory deprivation, and the other is on the workings of memory in the brain. The deprivation sequence allows Chuck Jones and company to have some real fun in the animation department. In the experiment, volunteers were put in soft beds, with hearing muffled and sight purposely blurred. Even arms were softly wrapped in bandages that dulled the sense of touch.

Deprived of sensory input, the brain loses touch with reality. Subjects described seeing strange visions such as marching pairs of glasses. This allows Warner's animation to have a field day, drawing various interpretations of these hallucinations. The marching glasses grow rubbery and distorted, and geometric shapes and little monsters fill the screen.

But perhaps the most interesting sequences in the film's last minutes deal with exploration of the human brain and memory. The work of neurosurgeon Dr. Wilder Penfield is highlighted, and Penfield himself shows up for a brief talk. For the record, his delivery is much less stilted than some of his other colleagues in the Bell series.

Dr. Penfield was one of the greatest neurosurgeons of the twentieth century, a man who was celebrated and honored in his own time. His passion was to unlock the mysteries of the human brain, and to alleviate suffering. He founded the Montreal Neurological Institute in 1934. Penfield, though born an American, became a Canadian citizen and was later described as the "greatest living Canadian."

Penfield also had some adventures in his life. He was bound for France, intending to work for the Red Cross in World War I, when his ship was torpedoed by a German submarine. He was injured, and he survived by sheer luck.

But some of his most notable work was his treatment of patients with epilepsy. Wilder reasoned that if he could locate the specific brain tissue that was causing the problem, he could remove or destroy it surgically. This pioneering technique was found to be successful in many cases. Part of the process involved stimulating parts of the brain with electricity.

But in developing this procedure for epilepsy, he stumbled upon an amazing discovery: a discovery that is fully showcased in *Gateways to the Mind.* Penfield's patients were under general anesthesia but were fully awake when their skulls were opened. By stimulating the temporal lobes—the lower parts of the brain on each side—Penfield produced vivid memories for the patients. These memories were colorful and much more "real" than simply recalling an event. It was as if the patient was reliving the incident again—sight, sound, and even smell.

Penfield had discovered that the brain is a recording device— even details unnoticed at the time are faithfully recorded. In *Gateways,* we hear actual patients talking during surgery. "Now I see them," one man says. "They laugh—I'm in a house in South Africa." A woman seems a bit amazed when she recounts "I hear a song—I haven't heard that song since high school."

Though the Warner Bros. films generally avoid the religious themes that Capra loved, there is a touch of the "divine" in *Gateways* final moments. Dr. Baxter summarizes what we have just seen, then adds that the senses provide us with knowledge. "How man uses this knowledge in shaping his world will determine the future of mankind."

But while Baxter speaks these final words, the screen fills with scenes from a concert—probably Easter services—in the famous Hollywood Bowl. As he is speaking, the swelling strains of Johann Sebastian Bach's *Dona Noblis Pacem,* an excerpt from his *Mass in B Minor,* fill the soundtrack. Baxter ends his comments just as the chorus sings a long drawn out "Amen." One suspects that the music was selected for its stirring, upbeat mood rather than its religious themes.

Gateways to the Mind aired on October 23, 1958. The critics were generally favorable, and it remains one of the more interesting and watchable of the series. One person who was not amused was Frank Capra. In fact, he probably didn't like any of the Warner Bell entries. There was a level of sour grapes in his assessments, true, but also the films were just not his style. Writing to Frank Baxter, Capra remarked, "What I miss most in the continuation of the series after I left were the poetic and spiritual overtones that radiate from the physical world around us."

The Alphabet Conspiracy was the next Warner production in the Bell series. Owen Crump produced the film, with Jack M. Warner as "Executive in Charge of Production." This is one of the last projects Jack "Junior" did for his mogul father. Jack Warner senior had divorced his first wife Irma, and Jack Jr. sided with his mother. Junior made no bones about disliking his stepmother Ann, and this led to an estrangement between the two men. Jack Jr. was fired, and he only heard of his dismissal by reading about it in the trade papers. He was even forbidden to enter the Warner lot in Burbank.

The screenplay used characters from the classic book *Alice in Wonderland* by Lewis Carroll, and was written by Leo Salkin and Robert Hobson. Salkin had worked for UPA on various cartoon shorts. *Alphabet Conspiracy* was directed by Robert B. Sinclair, a veteran who helmed many television shows.

Character voice work was done by Daws Butler, who, like Mel Blanc, was well known for his roster of impressions. Yogi Bear is probably the most famous of his memorable characters. The animated segments, many of them clever, were directed by Friz Freleng. He subsequently won an Academy Award for Best Animated Short Film in 1964 for *The Pink Phink*.

Alphabet Conspiracy's cast includes some well-known actors. Dr. Frank Baxter is on hand once again, still playing a generic scientist. This time, though, he's "Dr. Linguistics," a more specific area than merely "research." Since Baxter was a Professor of English at USC, this film comes closest to his actual area of expertise. When he talks about words and pronunciations, he can do so with real conviction and without acting.

Hans Conried plays the Mad Hatter in the film, complete with oversize top hat and bright yellow coat. Conried was one of those actors who one rarely sees anymore—a man of real versatility. He was noted for his precise diction and "Shakespearian" manner, though in *Alphabet Con-*

Baxter talks to Cheryl Callaway while the "Mad Hatter" (Hans Conreid) looks on with distain. Photo Courtesy Author's Collection.

spiracy he plays the role partly for laughs. Conried was born in 1917, and studied acting at Columbia University. He was a member of Orson Welles's famous Mercury Theater group, and did a lot of radio work in the 1940s.

Conried really hit his stride in the 1950s, appearing in many movies. He also made his Broadway debut in *Can-Can.* Baby boomers can remember his voice work, including "Snidley Whiplash" in the cartoon *Dudley Do-Right of the Mounties,* and the "slave of the mirror" on the Walt Disney television show.

The actor was seen many times on the big and small screen, essaying many diverse roles. Probably the most surreal picture he ever did was a bizarre fantasy called *The 5,000 Fingers of Dr. T.* Tommy Rettig plays "Bart Collins," a young boy who hates to take piano lessons from an autocratic Dr. Terwilliker (Conried). Bart dreams he is a prisoner along with 499 other boys and forced to play a giant piano.

The dream sequence is a musical fantasy with bizarre sets and almost "Busby Berkeley" action. The screenplay was written by Theodor Seuss Geisel, the famous "Dr. Seuss" of *The Cat in the Hat* and other classic children's books. The movie was nearly incomprehensible to audiences at the time, and it was a failure, but it has since attained the cult status.

The little girl in *Alphabet Conspiracy*, Judy, is played by Cheryl Callaway. She was one of those child actors whose careers flourished for a decade or so, and then they seemingly disappeared without a trace. Callaway's career seemed to end with her teenage years in the late 1960s, a common enough situation when child actors lose their youthful appeal. Born in 1946, in *Alphabet Conspiracy* she was twelve years old. Callaway was featured in many television shows in the 1950s—sometimes she was uncredited, or merely labeled "Little Girl."

The Korean War drama *The Bridges at Toko-Ri* was probably her biggest movie in terms of star power and budget. In the film she plays "Susie Brubaker," daughter of Grace Kelly and William Holden. In *Alphabet Conspiracy* she is charming, cute, and sweet without being saccharine. She's well cast in the film.

In one scene, the Mad Hatter invites Judy and Dr. Linguistics to a tea party. He doesn't want to socialize, however—he wants to trip "Dr. L" up by bringing in guests who will be hard to understand. The first guest is Shorty Rogers, a jazz trumpeter and composer. Rogers was a real personality, and he was famous in the "hip" music world. When Judy asks him how he is, Shorty responds with "Crazy."

But Baxter knows "beat talk", a lingo Shorty can readily understand. Dr. Baxter explains to Shorty that "The hat here (Mad Hatter) has been trying to bug me with his ABC jazz." Shorty is impressed with Dr. Linguistics's mastery of cool musician patter. He finally comments "you're a real live aspirin!" and exits with a flourish from his horn.

The next guest is a "theatrical publicity agent" who speaks a kind of "Broadwayese" that Daymon Runyon would have loved. He's played by Stanley Adams, a character actor who appeared in many films and television shows. He essayed "Cyrano Jones" in the "Trouble with Tribbles" episode of the classic *Star Trek*. Interestingly, Adams actually co-wrote another classic *Star Trek* episode: "The Mark of Gideon."

Sadly, Adams committed suicide in 1977 at the age of sixty-two. His back had been badly injured, and besides the obvious pain, he also probably felt such an injury would limit his roles. In *Alphabet Conspiracy* he's a fast-talking agent, his look not unlike Jackie Gleason in *The Honeymooners*. He shows Baxter his "lead" for an article—probably the performing "bible" *Variety*—and says "Tongues wag and Wags gag!" Baxter, unfazed, notes in the same dialect that Adams has "ballpointed" the situation nicely.

The third and final guest is a "southwestern cowboy" played by real-life cowboy actor Cactus Mack. A talented country-western musician, he also carved himself out a niche as a character actor. He was a cousin of fellow cowboy actor Rex Allen and Glenn Strange. Strange is best remembered for doing a turn as Frankenstein's monster.

Mack enjoyed a flourishing career in the 1950s and early '60s. He appeared in many classic TV westerns, including *Gunsmoke* and *Bonanza*. He died of a heart attack in 1962, only a few years after his *Alphabet Consipracy* outing. In his segment, he complains, "These doings give me buck fever, and I get hog tied when it comes to making chin music with strangers!" Once again, Baxter understands what the guest is saying, causing Mack to declare, "You savvy my lingo! You'll do to ride the river with!" Baxter, smiling, says to the Mad Hatter, "Him and me are made of the same leather!"

A moment later, Dr. Baxter explains that these guests are not speaking different languages as Hatter insists, but dialects. "These dialects and vocational vocabularies," Baxter explains, "are what make any language interesting and colorful. There is a science of American dialect geography, determining not just what people say, but how they say it."

This provides an opening for a very amusing cartoon segment. Two detectives are trying to determine exactly where a well-dressed suspect comes from. "We have ways of finding out," the lead detective declares. The detectives show the suspect a series of words on placards, and they demand he read them out loud. He does so, and with each set of words they narrow down his place of origin.

At one point, the detectives mistakenly show a placard upside down—and the suspect responds by standing on his head. By the time the suspect pronounces the word "greasy," the detectives have placed him within thirty miles of Times Square in New York. It's a very entertaining and easy-to-remember way of learning about dialect geography.

The film also shows early experiments that tried to get chimpanzees, our closest relative in the animal world, to form words and speak. There is some footage of Dr. Keith Hayes and his wife, who raised a chimp named Vicki in the 1940s. They tried to get Vicki to learn words, helping her by training and by manipulating her mouth. The results were poor. After much work she could only "say" four words.

There is one sad story attached to an otherwise entertaining and successful film. Jack Warner hired the great silent comedian Buster Keaton to do an extended sequence in *Alphabet Conspiracy*. It was going to be illustrating the concept of "nonverbal speech." The setting was a grocery store, and the skit allowed Keaton to use his skills in pantomime.

As the scene unfolds, Buster walks into the grocery story pushing an imaginary "grocery cart."

He walks around, carrying on nonverbal communications with all and sundry. Once his purchases (all unseen and imaginary) are gathered, he proceeds to the checkout stand, and he buys the watermelons, bottles, and bags of food. Arms loaded with the invisible goods, he walks out of the store and crosses a busy street.

Jack Warner Jr., who was on the set and watched Keaton from behind the camera, marveled at his artistry. The producer later recalled the comic was "a genius of improvisation, and he gave us exactly what we wanted. The script only included some suggestions and the entire action was left up to Keaton."

Warner went on to say that the script's brevity was "fortunate, as scripts have a way of smothering creativity—and in this case, we turned him loose and he gave us just want we wanted." When comic genius was finished, all were delighted with the footage. Everyone, Warner included, was sure that Keaton's segment would be a highlight of *Alphabet Conspiracy.*

Apparently it was a fairly extensive sequence, too, and one that fully showcased Keaton's talents. Finally, the finished show was completed and sent to AT&T and Ayer for final approval. Warner could not believe his ears when he heard the "suits" back in the board room utterly rejected Keaton's work. They argued that he slowed up the action, and that nobody remembered Buster Keaton anymore!

The collective stupidity of corporate minds prevented a good film from becoming a great film. Warner was aghast, but there was nothing he could do. After all, as he put it, AT&T was "paying the freight." Sadly and very reluctantly, the Buster Keaton segment was excised from *Alphabet Conspiracy.* It remained a bitter memory for Warner for years afterward. "We were all saddened by this decision," Warner recalled, "and the loss of what I recall was a gem of a performance."

The lost Buster Keaton footage has been something of a "holy grail" quest among film collectors and Keaton buffs. Warner himself wasn't sure of its ultimate fate. "I wonder if the film still exists," he wrote in 1993, "or if it molders in a forgotten vault, but I doubt it, as the Warner Bros. TV Division is long gone and has been succeeded by a whole new collection of companies using Time-Warner as its title."

The short sightedness and stupidity of the executives is incredible. As Warner put it, "Buster did it all beautifully—and the great crime was that the sequence was chopped out of the final cut because the deadhead old men of AT&T had forgotten their youth… So much for the leaders of American industry!"

After *Alphabet Conspiracy* Ayer apparently was dissatisfied the way Warner was handling the series. They offered the project to Frank Capra again, encouraging him to put in a bid. He wasn't interested, so Ayer and

Bell Telephone stuck with Warner Brothers. Warner did two more shows, *The Thread of Life* and *About Time.*

With *The Thread of Life* (1960) Warner tried a different way of telling the story. It's much less whimsical and humorous than its predecessors. Baxter is on a set with huge photo murals of human cells and the like. He's working alone, and he is noticeably thinner than he was in the previous Bell films. This is probably due to the fact that around this time he developed diabetes, and he had to really watch what he ate.

Before Baxter appears, there is a little prologue that sets up the premise of the show. Serious scientists in immaculate white lab coats are seen looking through microscopes and generally doing their research thing. It's a living example of the Bell Science credo, an effort to "study nature, and nature's laws." Chet Huntley, who is uncredited, does the beginning narration. Huntley is most famous for being the co-anchor of the NBC News *Huntley-Brinkley Report*. It was one of the greatest pairings in broadcast journalism.

David Brinkley, like Huntley, was a seasoned journalist but with more of a dry sense of humor and genuine wit. They were only linked electronically, since Huntley was in New York, and Brinkley was in Washington D.C. They became very well known, and entered into the realm of popular culture. Their iconic sign off—"Goodnight Chet, Goodnight David, and Goodnight from NBC News"—became a well-remembered memory of the sixties.

Dr. Baxter explains the double helix of DNA. Photo Courtesy Author's Collection.

The most distinguishing feature of the *Thread* set are six television screens arranged in two neat rows. Baxter walks over to the bank of TV sets early in the show. "Through the magic of electronics," Baxter explains, "we are inviting some of the audience to come along with us and join in." Of course, "some of the audience" implies immediacy, almost as if the program was live. This is all an illusion, of course, because the folks shown are from Central Casting not from the public at large.

One of Baxter's televised questioners is Don Grady, playing a "left handed boy." He later achieved fame as one of Fred MacMurray's kids in the sitcom *My Three Sons.* Another "TV head" is character actor Charles Seel, who appeared in many now-classic movies and television series. There are other actors on Baxter's televisions, most notably a man with "heavy brows," a woman with a "white forelock," a cute teenaged girl, and a nervous-looking couple.

Though the name of the woman is unknown, the white forelock actress allows Dr. Baxter to play the gallant and flirt a little bit. Smiling, he declares that her white forelock is very fashionable." He also jokes a bit with the other TV heads, at one point calling Grady a "chip off the old block."

Frank Baxter also has a good time poking a little fun at himself. The subject of male pattern baldness is brought up. Baxter says that both males and females have a baldness gene, but real baldness is not produced until the baldness gene is combined with testosterone, the male sex hormone. The doctor's eyes open wide for effect, and then he points to his bald pate and says "you get *this* eminent effect!"

Some commentators have theorized that the television screens were deliberate, in that they put a wall, a kind of distance between Baxter and his audience questioners. After all, this is the 1950s, and overt sex, even when discussed as science, was still taboo. There is an emotional distance in *Thread of Life* that mirrors the distance between Baxter and his questioners. The man's sperm and woman's egg "come together," and life is created. Granted, this is not a sex education film, but the presentation is very cut and dried.

"There are billions of people over the face of the earth," Baxter relates, "all sizes, shapes, and dispositions." His comments are accompanied by stock Warner "travelogue" style footage of peoples in Asia, Africa, the Middle East, and America. "How does nature endow us with so many different traits and characteristics?" Baxter asks.

The first real breakthrough in knowledge was the work of Gregor Mendel. An Austrian monk, he did pioneering studies of genetics in the late nineteenth century. The Mendel segment of the film is particularly well done, first with photos, a recreation of Mendel's pea plant garden, and finally with simple but effective animation. Mendel recognized "dominant" traits and "recessive" traits. This led science into the exploration of the units of heredity, later called genes.

"Man!" Don Grady exclaims, "And all without a microscope!" This boyish enthusiasm provides Baxter with an opening. "The most powerful instrument in the cause of science is the human mind," the good doctor firmly declares. Gone are the references to God that so sprinkled Capra's scripts. In *Thread*, the process of heredity is due to nature's laws and pure chance, not a plan "known only to God" as in *Hemo the Magnificent.*

Thread of Life also does a good job in explaining the chromosomes and mitosis—the way a cell divides. Experiments with fruit flies are shown, including segments on the bizarre mutations that occurred when the flies were subjected to radiation. The subject of radiation was a hot-button topic in the 1950s, due to the ever-present fear of thermonuclear war.

New York Times critic Jack Gould, who generally liked the Bell series, took *The Thread of Life* to task for its somewhat muted presentation of the subject. Gould recoiled at the images of irradiated fruit flies and their terrible deformities—some misshapen, blind, or without wings. A deformed ear of corn was also prominently displayed.

The 1950s and '60s was also the age of anxiety, when civilization, and perhaps life itself, seemed poised on the brink of nuclear self-destruction. Gould, very much a man of his time, says the program "inched closely to the phase of genetics having the greatest immediate newsworthiness- the consequences of radiation on human cells."

In the film, Baxter admits that "some" radiation effects are indeed harmful, but points to beneficial mutations, like hybrid corn and rust-resistant wheat. But in some viewers' minds the atomic "mushroom cloud" hovered over everything. The images of a nuclear Armageddon, with the survivors and their descendants deformed and misshapen, was not far from anyone's mind.

Gould felt that a matter of such supreme importance should have been handled more thoroughly. While nuclear power plant leaks are always a potential problem, today we simply don't share Gould's concerns. He goes on to say "An educational program that leaves unanswered the biggest question it raises hardly can be regarded as altogether successful."

The critic did seem to appreciate the fact that this Bell entry was less "sugar coated" than its predecessors. This seems to be an oblique slap at the earlier Warner films, and perhaps even the Capra editions. He considered less "sugar" as a positive development, but felt the television screens were somewhat off-putting.

The script is at great pains to underscore the fact that the issues of heredity and genetics are really "about us." They show us "how we are alike—in some ways—and yet why each person is different." And the key to human differences is in our DNA, a fairly new discovery at the time. Baxter walks over to a detailed model and gives a fairly interesting and understandable description of the famous double helix of a DNA molecule.

Baxter also gives due credit to the "discoverers" of the DNA strand, Francis Crick and James Watson. One was British, the other American. They are indeed the discoverers, and got a Nobel Prize in 1962 for their work. The film does not mention, however, that they were building on the work of others, notably Rosalind Franklin's high resolution images.

About Time is the last Warner Bell film, and the last in which Dr. Baxter makes an appearance. Produced and directed by Owen Crumb, *About Time* has a prologue that is somewhat reminiscent of the Capra titles. With fleecy clouds swirling in the background, an old-fashioned hourglass appears, and a narrator quotes the Bible, Ecclesiastes, 3.1: "It is written, to everything there is a season, and a time for every purpose under heaven."

Baxter appears and explains that, in order to better understand time and its concepts, we'll journey to an imaginary planet, Planet "Q." These aliens are humanoid and civilized, but they don't know much about time and would like to get things going by setting a giant prototype clock. Besides Baxter, there's the "King of Planet Q," played by Les Tremayne, and the king's "Butler," essayed by character actor Richard Deacon.

Tremayne was actually born in England, but came over to America at such an early age he was able to shed his British accent. He began playing in community theater and vaudeville shows. His deep, distinctive voice soon landed him jobs in radio. It is estimated that in the 1930s and 1940s Les worked on some 30,000 radio shows. The actor soon became a regular in movies and television shows. Perhaps his most memorable movie role was that of "General Mann" in George Pal's sci-fi classic *War of the Worlds*. He had a sixty-year career, dying at the age of ninety in 2003.

Richard Deacon was a character actor who usually played stuffy authority figures. The dignity and slight air of distain he conveyed made him a perfect butler. Deacon did achieve a measure of fame during his five-year stint as a character on the classic comedy *The Dick Van Dyke Show*. A gourmet chef in his personal life, he even hosted a cooking show in the 1980s. Deacon died of a heart attack in 1984. He was sixty-three years old.

As the plot unfolds, the Butler and the King ask Baxter questions about time. He responds by using a "telescope" to show how humans dealt with time and its measurements on earth. Baxter explains that humans first noticed the passage of time through the movement of the sun, moon, and stars. These heavenly bodies were gods to early peoples, because they seemed to determine the seasons. Once agriculture was invented, it was important to know when to plant and harvest.

As a result, calendars were gradually developed, but even the Gregorian calendar we use today is "slightly off." The problem, Baxter relates, is that a year is "365 days, 5 hours, 48 minutes, and 46 seconds." What to do about the excess time? Dr. Baxter also discusses the development of time pieces, including the chronometer in 1761, which allowed accurate navigation.

The audience is shown Galileo's first use of the pendulum in 1583, and also the use of quartz crystals and atomic energy for more and more accurate clocks. Some of the best sequences involve Albert Einstein and his theory of relativity. Both Carl Sagan and James Burke touched upon these subjects at one time or another in their programs, but *About Time* does a pretty good job of it even without modern special effects.

A cartoon segment shows how twin brothers age. One is an astronaut, and when he goes into space he travels the speed of light. He returns sixty earth years later, virtually unchanged, though his twin is now a very old man. That's because time is different in outer space, especially when travelling so fast.

About Time was telecast on February 5, 1962. About two weeks later, on February 20, astronaut John Glenn became the first American to circle the earth in a spacecraft. His *Friendship 7* capsule orbited the earth no less than three times. Times were changing, and they were changing rapidly. In 1956, when the first Bell Science show *Our Mr. Sun* made its debut, America didn't even have a space program yet. Explorer I, the first American satellite, was still two years into the future.

Perhaps it was inevitable that Ayer would finally turn to Walt Disney to produce what turned out to be the final Bell Science film, *The Restless Sea*. This production is the most obscure of the nine Bell films—so obscure it's hard to get solid information on it. There's a lot of misinformation too. Some references list *The Restless Sea* as an episode of the regular series *Walt Disney's Wonderful World of Color*, but that is not the case. NBC aired it on January 24, 1964, offering it as a separate Bell Science film, even if Walt Disney produced it.

There are two versions of *Restless Sea*. One is a fifty-seven-minute version, while the other is only a half an hour. The hour-long version was first aired on television, then converted to a two-reel 16mm film for classroom use. The half an hour version was also shown in schools, but there's no data on why Ayer, Bell, or Disney decided to have a long and short film released more or less at the same time.

The animation is superb, as might be expected from anything associated with the Disney "brand." The ocean floor sequences convey a mysterious and dark, almost surreal, feeling in audiences. Another clever segment features the animated letters "H" and "O." They merge and form a bicycle, as the soundtrack breaks into a lively jingle. "H-2-O, H-2-O, Two atoms of hydrogen, one atom of oxygen spell H-2-O. What do you get? Terribly wet!"

Many oceanographic topics are covered in a relatively short time. Among the topics discussed were ocean tides and the topography of the ocean floor. There is also extensive footage of various forms of marine life, including sharks. Fact after fact is laid out in a fairly entertaining manner, explaining that the sounds that fish make are amazingly varied, or that the H.M.S Challenger undertook a three-and-a-half-year cruise in 1872 that was the first truly scientific study of the ocean.

Since he is the central animated character, much of the film's effectiveness is dependent on whether you like Sterling Holloway or not. It must be admitted Holloway's vocalizations are an acquired taste. For some, his voice grates on the ear; one observer likened it to "Brillo on sandpaper."

Apparently in the longer, hour version, Walt Disney himself plays host, much as he did in his popular TV series. In the half-hour edition he's nowhere to be seen. The only other "cast" member is a "drop of water" voiced by Sterling Holloway. Holloway, it will be recalled, appeared years earlier in *Hemo the Magnificent.*

On the surface at least, Disney seemed a natural choice to produce a science documentary. In the 1950s he was famous for his "True-Life" nature adventures, discussed with more detail in another chapter. Somehow *The Restless Sea* didn't quite make the Disney canon of memorable films. John Marshall Cuno, a reviewer for the *Christian Science Monitor*, gave *Restless Sea* a generally good review, and added, "The program is the last in the original series of nine beginning back as far as 1956. One hopes there will be a new series (the scientific advisory board is considering the possibility.)"

We don't know where Cuno got his information, but in the end it was Ayer and Bell, not any "advisory board," who would ultimately decide if the series would continue. Apparently Disney was the last hope. Ayer was not pleased with the Disney offering, so the series ended in 1964. But the eight earlier films enjoyed a continued life in American classrooms for another twenty years and more. The Bell Science series remains one of the milestones of educational television and film.

8 After Baxter: James Burke and Carl Sagan

FIFTEEN YEARS AFTER the last Bell Science program aired, new techniques and a new vision further transformed how science and technology were presented on television. The first of these was *Connections: An Alternative View of Change* (1978), a British Broadcasting Corporation (BBC) production that was aired in America a year later. James Burke was the writer and presenter of the ten-part series, a man very much in the Baxter mold in terms of charisma, and with an uncanny ability to explain complex topics in a very accessible way.

Burke opened up the science documentary by extensive location filming in far-flung corners of the globe. He also used models (a Baxter technique) and historical recreations with costumed actors. The BBC was no stranger to location filming, using it extensively in Kenneth Clark's *Civilization* and Jacob Bronowski's *The Ascent of Man*. But Burke added a new visual twist by doing things like riding in a hot air balloon, or gliding along in a snowmobile.

When *Cosmos* aired on PBS in 1980, the final element was added that brought the TV science documentary to full maturity: computer generated imagery. Dr. Carl Sagan, an astronomer and astrophysicist at Cornell University, was the writer and presenter of the series. Sagan, too, had a level of genuine charm, but his charisma was more laid back than Baxter and less "zany" than Burke.

James Burke was born in Londonderry, Northern Ireland, on December 22, 1936. Son of a businessman, he had no family academic tradition. Burke went to Maidstone Grammar School in Kent, which even today is considered one of the best schools in England. He attended Jesus

College, Oxford, focusing on the historical development of the English language. As a scholar, Burke seemed to be destined for an academic career. Destiny, or he might say serendipity, decided otherwise.

Armed with a newly minted Master's Degree in English, young Burke pounded the pavement—or occupied the phone booth—in search of a teaching position. He was on the phone with a Greek college when a friend told him of a job in Bologna, Italy. His imagination was fired. The land of ancient Rome and the Renaissance! According to some stories, he dropped everything—including the phone receiver—and contacted Bologna at once.

Burke got the job, and he soon left the foggy shores of dear old England for sunny Italy. He felt, he once said, like he had "died and gone to heaven." Burke had a solid foundation in the classics and in the Renaissance, so it was a thrill to actually live where much of western culture was born.

Once settled in Bologna, Burke found himself teaching English to "bored Italian graduate students." Soon he became bored too. Bologna is the source of the word "baloney," and he was starting to feel he was contributing to the definition with each lecture. The fledgling scholar did have a knack for languages, picking up Italian with relative ease. Unfortunately, the first Italian he learned was with the Bolognese dialect, which was almost unintelligible to other Italians. When he visited Rome and tried out his new language skills, the Romans laughed themselves silly.

Burke tried other things, including the creation of an Italian-English dictionary. He became a very good linguist, mastering Italian, Middle English, and French. Judging from his programs, Burke can also spring into Old English at the drop of a drinking horn. Perhaps his most unusual job was interpreting at the Vatican II—the great Catholic Church Council. He was working for B'nai B'rith, a fact he finds somewhat amusing today. "I was a lapsed Protestant interpreting for the Jews at a Catholic conference!"

But, once again, boredom threatened, when he heard that a British TV company was looking for someone to produce "bits" from the Mediterranean region for them, he responded with alacrity. He knew absolutely nothing about television work, but by keeping "one step ahead of the sheriff," as he once put it, he learned all he could while doing it.

Finally, Burke was recruited by the BBC and returned to England in 1966. Burke was still more of a scholar and teacher than television professional, but paradoxically that was a plus, not a minus, on his resumé.

Burke was soon in the new Science and Features Department of the BBC. As he recalled, lack of expertise was actually a plus. In an interview for Indiana Public Media in 2009, Burke explained, "The BBC recruited only humanities graduates for science programming on the basis of 'if you understand it, they (the public) will understand it.'"

The British broadcaster became famous in his home country by being the BBC science reporter and anchor who covered the Apollo manned missions to the moon. It's not that Burke lobbied for the job—in fact, he says today, "I was conned into it." His producer asked Burke what he knew about rockets. "Well," he said glibly, "there's a point at one end, and a fire in the other!" "You're in deep doodoo," the producer replied.

The producer explained that Burke was going to cover the Apollo missions, like it or not. In a meeting with higher ups, the producer had praised Burke to the skies, saying he "wrote the book" on rocketry. Not that it mattered, because Burke could "read the book"—i.e. bone up on the facts—before he flew to Cape Kennedy. The producer's instincts proved right; Burke was a natural for such an assignment.

James Burke unknowingly moved one step closer to "*Connections*" when he was asked to join a weekly magazine format show called *Tomorrow's World*, made up of five-minute items on technology. He started to realize that the pace of change was accelerating, making humans more and more dependent on modern technology. But, by the same token, the general public had little or no understanding of the very gadgets that were irrevocably changing their world

Taking a bit of inspiration from Winston Churchill's famous quote on the Royal Air Force, Burke summed it up by saying "never have so many people understood so little about so much." But if he could explain this new vision of history, science, and technology in a colorful, engaging manner, people *could* at least start to understand what was going on. Or, at least be made more aware. "Come on," Burke wrote, "you can do it—you don't need a Ph.D. You have 82 billion neurons in your brain, just like Einstein."

The idea of *Connections* was germinating, but it didn't truly sprout until Burke noticed a footnote in a book called *Medieval Technology and Social Change* by Professor Lynn T. White. The footnote mentioned that, in essence, the knight's stirrup allowed him to stay in the saddle and made him a more effective soldier. Feudalism was based on a hierarchy of mounted warriors who owed allegiance and military service to the man "above" him. Thus, the stirrup was the foundation of feudal society.

Well and good, but Burke took this footnote a step further. Because the Normans had the stirrup, they defeated the Anglo-Saxons at the Battle of Hastings in 1066. All of British history was altered—most notably our language. If the Anglo-Saxons had *not* been defeated, English would not have become half-French, which it is today. Ever the linguist, Burke declared that if the Anglo-Saxons had the stirrup, we might be saying "a fenchyscan ahton waelstowe gerweald," or "the French won."

The BBC gave Burke the green light for *Connections*. Money was plentiful, but Burke says with tongue-in-cheek that they were "giving you enough rope to hang yourself." If the project was a hit, well and good, if a failure, "your body would be found at the bottom of a well."

In *Connections,* James Burke argues that the origins of the modern world should be studied in a more comprehensive, interdisciplinary way, because nothing occurred—and this is particularly true of technology—in splendid isolation. Most history has been taught in a linear way—a succession of kings, or battles, and so on—because it's easier to teach in classrooms.

But Burke contends that everything is connected, and great inventions have largely been the product of chance and the whole gamut of human emotions, such as hope, ambition, greed, idealism, and so on. The paradox of history and technological change is that it's predictably unpredictable.

Burke also rejects the notion that one genius is solely responsible for progress, i.e. the "father" of this and the "father" of that. Take, for example, James Watt, the "father of the steam engine" according to the textbooks. Thomas Newcomen created a kind of steam engine, helped by the cheap and durable iron Abraham Darby created with coke. Watt improved Newcomen's design, and John Wilkinson's cannon-boring machine made strong cylinders.

"In spite of the myth," Burke says in *Thunder in the Skies*, "James Watt did not invent the steam engine. He just invented a vital bit of it, and its only when all the bits come together, that's when the final form of the steam engine comes together. And when it does, that's when widespread change occurs." In this case, the change was the Industrial Revolution.

Connections includes such entries as *The Trigger Effect, Death in the Morning, Distant Voices, Faith in Numbers, The Wheel of Fortune, Thunder in the Skies, The Long Chain, Eat, Drink, and Be Merry, Countdown,* and the summation of the series, *Yesterday, Tomorrow, and You.* In each episode, Burke usually starts with a modern invention—say, a 747 jumbo

James Burke in *Connections* (1978). Photo courtesy James Burke.

cargo plane—then goes back in history to see its origins. But while doing so, he illustrates the interconnectivity of innovations and events along the way.

In *Eat, Drink, and Be Merry*, Burke starts with a credit card and the concept of credit, then goes to the fifteenth century and the Dukes of Burgundy. They borrowed heavily, but the last Duke was defeated by the Swiss, who used pikes. Pikes were replaced by muskets, and armies grew larger. They had to be fed, and by Napoleon's time bottled and canned food was in vogue.

Food spoiled, and people blamed "bad air, which was called malaria." That is where the disease malaria gets its name. Investigations in "bad air" led to refrigeration, and the thermos, which could hold hot or cold liquids, was invented. Robert Goddard and others took the thermos concept to build rockets full of hydrogen and oxygen fuel. This led to the German V-2 rocket of World War II, and ultimately The Saturn V rocket that took us to the moon. So the humble plastic credit card took us to the moon.

James Burke's concept of history and the development of science and technology was indeed, as the subtitle plainly states, "An alternate view." How could he get the average viewer interested? How could he make this material accessible, and even enjoyable, to the general public? A tall order, and one that the two Franks—Baxter and Capra—had wrestled with twenty-five years before.

Burke came up with the same kind of solution: first, he provided the show with colorful visuals. The Bell series had colorful cartoons, models, some acting vignettes, graphics, and film clips. Burke took this a step further and added full-blown historical reenactments and location filming across several continents.

He also recognized that you had to make the material viewer-friendly. Speaking from a British perspective—he probably didn't know about Baxter and the Bell Science series—Burke said in an interview, "you had to break the mold. Up to that time, people (presenters) were stuffy, spouting polysyllabic gobbledygook. A lot of the people out there found the material threatening: 'important academic' stuff."

In Burke's view: "you have to make the material accessible—have people drop their guard. So that's why I didn't wear a suit, and did rather silly things." In fact, the presenter's leisure suit—white, with a brown shirt—became a kind of signature costume. With the passage of time, however, this asset became a dated '70s curiosity. Today, when asked what became of the leisure suit, he responds smilingly, "I burned it."

Burke also discovered, as Baxter and Capra before him, the value of humor. "Humor is a tremendous aid in getting an idea across," he asserts, "because if they laugh at something, they don't feel threatened by it, that 'I'm not too stupid to understand this.' Humor is a great door to get people into the material."

The Long Chain is one of the funniest of the series, and best illustrates what Burke is talking about. In one segment, he speaks of a German with the (Anglicized) name of Fred Albert Windsor. He was "the flashy, the incompetent, the semi-illiterate" who did help provide nineteenth century Britain with gas light. In selling shares for his company, he claimed you could "put 50 pounds in and [get] 6,000 pounds out." Burke wryly comments that the only thing his listeners got out of the affair was "a headache from the escaping gas jets."

Later on, Burke speaks of a man named Perkins, who gave Great Britain a head start in the artificial color industry. But the British preferred to put money into colonial projects, not the chemical color industry. The Germans became the chief producers of color. Producing a large handkerchief, he covers his nose, gives a big "honk," and says "They blew it!"

But the funniest moment comes when Burke, seated at a piano, explains that the Germans and their colors became "dyed in the wool." Some German composers even wrote "colorful" tunes. Burke says, "Now here's a catchy little number that I think will have your feet tapping, or fingers itching, or head throbbing or something…" Burke tickles the ivories, playing a song named *Indigo*, while an earnest chorus of middle-aged Germans provides a robust, slight off-key accompaniment. "I can't think,' Burke wryly comments, "why this tune isn't better known in the concert repertoire. Or maybe I can."

Connections was an enormous undertaking, involving filming all over the world. It also was time consuming involving weeks of travel. "There are two minutes of edited film for each shooting day," Burke once explained. "In a ten-hour series, there's 600 minutes. Unless you want to do the series for the rest of your life, you have two film crews in the field. I remember taking two Concorde (supersonic) jets back to back—one from Kuwait to London and then the next from London to New York."

The project schedule was derailed by underwear. Well, almost. After days of intense effort, a shooting schedule was put together, only to be nixed by a savvy production assistant who said "You don't have knicker breaks." The production assistant said that Burke and the crew would be on the road for weeks—who'd do the laundry? They would have to halt

James Burke today. Photo courtesy James Burke.

and stay at a hotel for at least two days, or the hotel would not wash their clothes. "Knicker breaks" were quickly added to the schedule.

Burke had several adventures during the course of the film shoot. The worst he can recall—or chooses to relate—was "changing clothes in a Malaysian train station toilet." On another occasion, he was deposited on a glacier in Antarctica. The idea was to have a kind of dramatic overhead shot with Burke on the ground in the distance, a tiny "spec" in a huge expanse of white. Something, in fact, a little like the celebrated opening shot of Julie Andrews in *The Sound of Music*.

The camera helicopter left for what was supposed to be a very brief time. The minutes ticked by, and still no helicopter. It looked as if Burke was abandoned in the middle of a frozen wilderness. After twenty nerve-wracking minutes, Burke recalls "I was wondering if I should have changed my career." Luckily, the chopper returned to the scene, the pilot having briefly been lost himself.

Connections has a rich, big-budget feel, especially with the historical reconstructions. True, the BBC was generous with funds, but still, there were limitations. But Burke says they didn't actually pay for much. He recalls that in one segment, he wanted to discuss economic development in the Middle Ages. "Boring!" Burke himself declared. But it seems there was a kind of a medieval-theme festival with costumed reenactors somewhere in Europe. They agreed to be filmed as long as they had "all the beer they could drink." Problem solved!

Burke's *Connections* was an enormous success, and in 1979, when it was aired in the United States, it was the most watched PBS program up to that time. It aired in fifty countries, and was part of the curricula of some 350 colleges and universities around the world. *Connections* was such a success it spawned sequels, such as *Connections 2* (1994) and *Connections 3* (1997).

These later shows are well produced, but occasionally Burke—or his researchers—make connections that are dubious and sometimes downright false. In "Déjà Vu," he mentions the cochineal insect, which produced a scarlet dye that colored British "redcoat" uniforms. As the visuals show Queen Elizabeth II reviewing her guards, Burke's voiceover informs us that a "distant ancestor" of the Queen, Frederick the Great, practically created the modern disciplined army. That's why he "never lost a battle."

Well, interesting, except none of these statements is true. Frederick the Great died childless—he had no descendants. (His family is connected to the Windsors, however.) His father Frederick William "the Prussian Drill Sergeant" was the one who created the army. Frederick the Great was defeated on a number of occasions, especially at Kunersdorf in 1759. But these are very minor quibbles. When you are dealing with projects like the different *Connections* series, with far-flung locations and a multitude of subjects, minor errors are bound to creep in.

Connections and its sequels were a major landmark in television science programming. Burke has gone on to start developing the James Burke Knowledge Web. It's a logical extension of his many years of work. It's a work in progress, moved along by the dedicated work of volunteers. Its stated purpose is, as the website proclaims, to have a place "where students, teachers, and other knowledge seekers can explore information in a highly connected, holistic way that allows for an almost infinite number of paths to explore among people, places, things and events."

Cosmos: A Personal Journey was the next, and in some respects final, phase in the development of the educational/science documentary. This series was the first to showcase CG visuals in a big way, producing such wonders as host/presenter Sagan "visiting" the ancient library of Alexandria. Audiences were dazzled, and such techniques soon became standard.

Carl Sagan was born on November 9, 1934, in Brooklyn, New York. His father, Samuel Sagan, was a typical Russian Jewish immigrant, seeking asylum from a Czarist world where anti-Semitism and terrible pogroms were the norm. He was a garment worker, who was later awed by his son's range of interests and intellectual brilliance. Sagan's mother Rachael had intellectual ambitions of her own as a young woman, but the social restrictions of the time—women were expected to be wives and mothers—coupled by her Jewish ethnicity, frustrated such desires.

Instead she in a sense lived through her son Carl, since he would fill her unfulfilled dreams. And indeed, young Carl showed promise at a very early age. He was fascinated by dinosaurs, like most children even today, and visits to New York's American Museum of Natural History left him enthralled.

A budding genius, Sagan also had a sense of curiosity coupled with a sense of awe. Many times, young Carl would spend nights looking up at the star-speckled, inky void. A library book revealed a startling fact: The sun was really a star! And those little pinpricks of light were really suns, but too far away to see clearly! "The scale of the Universe opened up to me. It was a kind of religious experience. There was a magnificence to it, a grandeur, a scale which has never left me. Never ever left me."

But the really big turning point in Sagan's life seems to be when he was about five, and the family visited the celebrated 1939 New York World's Fair. Here was a heady glimpse of the future. Sagan recalled, years later, that exhibits like General Motor's World of Tomorrow fired his imagination. "It showed beautiful highways and cloverleaves and little General Motors cars all carrying people to skyscrapers, buildings with lovely spires, flying buttresses—and it looked great!"

Other World's Fair wonders included talking robots, photoelectric cells, and a curious new device called television. There was also the burying of a time capsule containing relics of 1930s America for the people of the distant future to see. Sagan loved the idea of a time capsule.

Sagan became a full professor at Cornell University in 1971, and was the Associate Director of the Center for Radio Physics and Space Research. He also made contributions to the scientific field. He theorized, for example, that Jupiter's moon Europa might have subsurface water on it. If it had water, it might then be habitable for humans. His theory became fact when the spacecraft Galileo did indeed detect liquid compounds where he said they would be.

Dr. Sagan also became a consultant for America's budding space program from the 1950s onwards. In fact, he briefed Apollo astronauts before their famous journeys to the moon. But Sagan is perhaps most identified with a series of robotic spacecraft that were launched to probe the outer limits of the solar system. *Voyager 1* and *Voyager 2* were launched in 1977, and it was said that they more than quadrupled our knowledge of the planets of our solar system.

Carl Sagan was a firm believer in extra-terrestrial civilizations, and so he was instrumental in having discs carried by *Pioneer 10* (1972) and *Pioneer 11* (1973) that carry pictorial messages from the planet Earth. Included are a naked man and woman (very PG rated), the man raising his arm in greeting. But even more sophisticated messages were place aboard *Voyager 1* and *Voyager 2* in 1977.

Each spacecraft had a "golden record" which showcased the Earth and its human populations for any alien civilizations that might find it. There were spoken greetings in fifty-nine languages, sounds of nature, and pictures of men, women, and children from throughout the world. Sagan chaired the committee that selected the golden record's contents.

Cosmos was a coproduction of the BBC and Los Angeles affiliate KCET, with a budget of some $6 million for making the series and $2 million for promotion. In fact, *Cosmos* had the BBC's "fingerprints" all over it, including dramatic reenactments and stunning visuals. It was said that *Cosmos* used videotape for interiors, switching to film only for exteriors.

The thirteen-part series was noted for its dreamy, almost ethereal music segments, most notably from the Greek composer Vangelis. The last movement of his *Heaven and Hell, Part I* became *Cosmos*'s haunting signature theme. In fact, the enormous success of the series also introduced millions to the composer's works.

Among topics explored in *Cosmos* are: the origins of life; the search for life on Mars; the lives of stars; interstellar travel; and the danger of human civilization destroying itself though pollution, overpopulation, and other ills. Even in the 1980s he worried about the greenhouse effect and global warming. Sagan also explored the past, dating back to the time of the ancient Greeks. Scientific heroes like Copernicus, Tycho Brahe, and Johannes Kepler are profiled.

Cosmos titles are in keeping with the dreamy sense of wonder that Sagan sometimes imparts during the course of the series. There's *The Shores of the Cosmic Ocean, One Voice in the Cosmic Fugue, Harmony of the Worlds, Heaven and Hell, Blues for a Red Planet, Traveler's Tales, The Backbone of Night, Journeys in Space and Time, Forever, The Persistence of Memory, Encyclopedia Galactica, The Lives of the Stars, The Edge of Forever,* and *Who Speaks for the Earth*?

The last episode, *Who Speaks for the Earth*, is perhaps Sagan's most personal and compelling entry in the series. He basically says that humans are at the crossroads—we now have the power to destroy not only civilization but the earth itself. This danger weighed heavily on Sagan's mind, particularly the threat of nuclear war. But he also offers some hope that we can escape, even solve, problems that are largely of our own doing.

While charming and personable, Sagan is much less humorous than Baxter and Burke. He's pretty straight forward and serious, though his somewhat "professorial" demeanor in no way lessens the overall effectiveness of the series. As Jim Blinn admits, "Carl Sagan was a nice man, but at times he sounded more pompous than he meant to be. The producer did a nice job of editing his more pretentious pronouncements, so he comes off sounding like a down to earth guy."

The only "signature" humor came from Sagan's speech rhythms in pronouncing words. He became linked to a catchphrase "billions and billions," which he never actually said in the series. He did emphasize the "b" in billions, in order to make sure the audience didn't mistake it for "millions." The catchphrase became the source of many comedy routines.

On a slightly negative note, Sagan's media fame was rumored to have been a factor why the National Academy of Sciences denied his membership. Frank Baxter, too, was unpopular with many of his USC colleagues, who jealously dubbed him "the Liberace of the Library." But, like Baxter, Sagan seemed to have brushed off their comments with a good-natured smile.

There's no doubt that Sagan relished his fame. But that was not his primary motive for such efforts as *Cosmos*. The professor felt that it was in the interest of the scientific world to inform the public. After all, in the end, much of the funding scientific research gets—or used to get, with recent budget cuts—is from the government. And so the money comes from the public in the form of taxes.

Sagan was also a controversial figure, at least in some quarters. A self-professed agnostic and "secular humanist," he provoked the ire of right-wing religious groups. He maintained that God cannot be proven or disproven by science. Sagan didn't believe in a white-bearded patriarch up in the sky who counts the fall of every sparrow. On the other hand, if God is the sum total of the physical laws of the universe, that kind of deity is emotionally unsatisfying. After all, as he once put it, it's difficult to pray to the law of gravity!

But the *Cosmos* series was revolutionary in its use of CG visuals. James Blinn was the man most responsible for the dazzling CG effects that highlight the programs. Blinn, a brilliant young computer artist, was working for NASA's Jet Propulsion Laboratory at the time. When he heard that PBS station KCET was planning the series, he became immediately interested. This was right up his alley; earlier, he had created some wonderful images for NASA of the *Voyager* spacecraft as it flew by a bright-ringed Saturn and other planets.

Jim Blinn created a demo to show Sagan and the producers. They were impressed, and Blinn and his team were hired. The CG visuals that he produced more than lived up to expectations. As computer scientist Ivan Sutherland once remarked, "There are about a dozen great computer graphics people in the world total, and Jim Blinn is six of them."

Blinn admits today that he was profoundly influenced by Baxter and the Bell films, as he noted in a recent interview:

> *I first saw the Bell series when I was about eight years old. I especially remember* Our Mr. Sun, Hemo the Magnificent, *and* Meteora: The Unchained Goddess. *Each of these had a different supporting cast, but they all had in common Frank Baxter as the main host. What I remember about him was his reassuring presence. He gave the impression of being knowledgeable without being condescending or overly "show-offy." He just had a joy in sharing his knowledge with others. He became the "face" of science, and it was amazing to find in real life he wasn't a scientist but an English professor.*

Thinking about it today, Blinn admits that:

> *The animation also impressed and engaged me. I liked that the character interactions gave the impression that there was more going on that wasn't covered in the movie, more to learn elsewhere. I recall the magician in* Our Mr. Sun *pointing his wand at the bottle of hydrogen and yelling "Fusion!", and the small characters in the* Unchained Goddess *representing Snow, Sleet and Hail sprinkling out their respective precipitations from bowls.*

In fact, he was so inspired that: "Later, a high school friend and I tried to do animation using paper cut-outs and an 8mm movie camera. Even then, I wanted to make something educational. One of our efforts was about the development of the wheel, and very similar to the Bell Science series."

Jim Blinn believes that seeing the Bell films was so universal from the 1950s to the 1970s they united generations otherwise different in location, race, ethnic background, or class. A kind of collective memory made even strangers "know" each other. Blinn cites his own experiences with his wife Amanda.

> *I can still remember the scene in* Hemo the Magnificent *of the three little men with levers showing how blood can go to the brain or to the stomach or to the body. My wife, Amada, saw* Hemo *when she was a kid and that image stuck with her too. Through those films, people who grew up in different places had shared experiences. If they met later on in life, they weren't completely starting from scratch in a relationship.*

Blinn says today, "Most of the CG I did for the series was simple black and white line drawings showing Kepler's laws, constellation evolution with star proper evolutions, etc. I did several color scenes of the *Voyager* spacecraft mission which were adapted from the *Voyager* fly-by animation I did for the Jet Propulsion Laboratory."

The other big main scene he did for *Cosmos* was a color animation of DNA replication which he admits "was a killer to do, but it came off well when finished." But he certainly was pleased with the finished product.

And so Blinn makes the story come full circle by providing a tangible link to Baxter and Bell science. If it wasn't for Baxter and Bell Science, *Cosmos* might not have even been created, or at least would have been vastly different. Then too, if it wasn't for Baxter, perhaps there wouldn't even be a PBS to air such shows as *Cosmos*.

Blinn was eager to contribute to *Cosmos*, but there were a few potential snags to work out: he was, and continued to be, a full time employee of NASA's Jet Propulsion Laboratory (JPL). It took some sensitive negotiating to persuade the government-sponsored lab to sign a contract with a television production company, even if that production was going to be aired on PBS. But all parties agreed, and Blinn and his team started work.

Probably the most talked-about scene in the series featured Carl Sagan realistically walking around the Great Alexandrian Library circa the first century B.C. It was a seamless creation, realistic in every detail, even though Sagan was actually walking around a green-screen stage. The actual "Alexandrian Library" was a scale model, not computer generated effects.

Another segment featured Sagan with a look of rapture on his face (he was an advocate of marijuana!) piloting the "Star Ship of the Imagination." While the good doctor fiddles with a control or two, he watches a screen very much like the one on the Starship Enterprise of *Star Trek* fame. As Sagan journeys through space, he gazes in wonder at supernovae, gaseous nebulae, and other fantastic sights.

With *Cosmos*, the television science documentary came to full maturity. It was a slow process: a thirty-year evolution that can trace its lineage back to the Bell films and Dr. Frank Baxter's unique and charismatic teaching style.

9
All's Well That Ends Well

Dr. Frank Baxter retired in May 1961, after teaching at USC for thirty-one years. He remained, as he admitted in a letter, "busier than ever." In fact, the professor was a key element in an imported Shakespeare series that was public television's first really major hit. But after about 1970, the jobs became fewer and he drifted off into relative obscurity. It was relative because the Bell films were still being shown in many schools, and his memory was green in the minds of the thousands of former students.

Baxter died in 1982, but the popularizing of science and technology on television continued to develop, a process that goes on to this day. James Blinn, who worked on *Cosmos* and was influenced by the Bell films, went on to be a major contributor to *The Mechanical Universe* in the mid-1980s. It's a project that he's particularly proud of, and with good reason.

William Sanford "Bill" Nye is another popular science educator who is best known for his PBS series *Bill Nye the Science Guy*. The program aired from 1993 to 1997 and garnered several Emmys. A man of many talents, he is not only a scientist but also an actor, writer, articulate defender of science in commentaries and talk shows, and even a dancer. In 2013, he was a contestant on the popular television show *Dancing with the Stars*. Unfortunately a leg injury hampered his performance and he was eliminated.

Cosmos: A Space-Time Odyssey was one of the latest efforts to take science to the general public when it was aired in 2014. As the title implies, it is essentially a reboot of the original Carl Sagan series. Ann

Druyan was one of the prime movers of the new series, which was hosted by astrophysicist Neil deGrasse Tyson. Druyan, who is Carl Sagan's widow, was also his collaborator on a number of projects. In fact, she co-wrote the original 1980 series. After twenty-five years, Druyan felt it was high time to revisit the concept.

In a 2013 interview conducted at the San Diego Comic-Con, Druyan explained that, "Carl Sagan taught me respect for the audience and the public, and there is no need to dumb anything down. Just speak clearly, and use the words we all use to depict the grandeur of nature." This is exactly how the Bell Science films presented their material in an effort to "understand nature, and nature's laws."

There are animated sequences in *Cosmos: A Space-Time Odyssey* that detail scientific discoveries of the past. They were produced by animator Seth MacFarlane, best known for his television cartoon series. The series also featured stare-of the-art computer graphics of the sun and other heavenly bodies. It seems the basic techniques that Frank Capra pioneered are still flourishing sixty years after their debut.

Popular science reached a whole new plateau when the Science Channel was launched in 1996. *Through the Wormhole*, a Science Channel offering, is the latest incarnation of what Baxter and Bell Science launched sixty years before.

When Frank Baxter announced his retirement in the spring of 1961, it was as if an atom bomb had been detonated on the University of Southern California campus. He turned sixty-five that year, and as he explained, "the University very kindly was willing to have me stay on for a few more years, but it seemed best to go." He added, "I have been very busy with television and radio, and there seems to be no end to the speeches and talks I could make if once I let down the floodgates. It looks as if I will be busy as long as I can stand up and make noises."

Baxter's last class in Shakespeare was literally a media event, extensively covered by the press. The subject was the Bard's *Twelfth Night*, and the professor didn't disappoint, regaling his audience with the usual acting out, entertaining asides, and bit of trivia. But everyone was waiting for the last few moments of his lecture, when he would drop the scholarly façade and talk to the fifty students on a personal level.

The last class took place in room 226 in Founder's Hall, which was packed with students as usual. Concluding his lecture, Baxter said, "I say farewell to you at this point." But the students were not fooled. They knew this was just a pause, a shifting of gears from the scholarly to his own vision and philosophy of life. It was a philosophy that was tinged with a certain wistfulness and regret.

Baxter was born into the late Victorian period, and he grew up around the turn of the century. When he was growing up, there was a kind of optimism about America and the world, in spite of the poverty and occasional social injustice. Even though by the 1950s we had progressed materially, and certainly lived better than our forebears, it came at a price. It was a price that Baxter felt was increasing anxiety in our everyday lives. The professor keenly felt how the threat of nuclear annihilation seemed to cast a pall over all human endeavors.

"You're in for trouble," he told his students in that last class. "You will know no peace, and your children will know no peace. The most you can hope for is Cold War. There is no sweet agreement possible. How vulnerable we all are—with our little flutter between the darkness and the darkness."

Yet Baxter's philosophy wasn't all pessimism. Humans could still learn from mistakes and perhaps avoid disaster. But to the professor, the way to gain knowledge was reading. It was a constant theme in his lectures, and he practiced what he preached. "We are not only a body and a bowel. People who bring children into this world and are not prepared to feed their brains are, in my philosophy, ignoble. You can't live, you can't mesh with this world, unless you *read*."

Ironically, Dr. Baxter couldn't help taking a pot shot at television, which some might think was biting the hand that "fed" him. "The idiots who run TV," he opined, "and look at the ratings, think people are best pleased with the low, hypnotic, and opiate level."

There was a final epilogue before he took his gold watch from the lectern—a standard flourish—and called it a day and a career. "You take the next part of this course," he said with a smile, "or I shall *haunt* you." Baxter then bowed and left with a warm and thunderous applause ringing in his ears.

Baxter kept in contact with Frank Capra for a few more years. In a letter dated November 13, 1961, the professor writes "And it still goes on! Mr. Sun continues to shine on still another school generation." When you add up the numbers of students seeing the Bell shows each year, the "score becomes astronomical. Certainly, the telephone boys more than got their money's worth as a result of this noble experiment."

In the same letter, Baxter notes that Capra has returned to feature films. In a reference to Capra's 1961 movie *Pocket Full of Miracles,* Baxter says, "Like everybody else I am delighted that you are still working your magic miracles. After seeing some recent movies, and reading some highly touted Broadway plays, I am convinced that the old sure fire formula of 'Lincoln's Doctor's Dog' no longer is workable. It is my feeling that the new formula is 'Lincoln's Doctor's Dog's Homosexual Drug-Addicted Psychiatrist'"

This was in 1961, when censorship was weakening but still largely in full force. One wonders what he would think of twenty-first century plays and movies. Again, Baxter was an older man who understandably could not relate to the "modern" trends.

A couple of months after his official departure, in July 1961, the newly retired Baxter was going to help rescue public television for the second time. In the early 1960s, public television found itself in the doldrums—the wind taken out of its sails by the overwhelming popularity of its commercial rivals. Baxter's *Shakespeare on TV* apart, too many public station offerings were still dry and pedantic. It had begun life as a kind of video classroom, and, try as it might, it was still largely stuck in that mode.

The public broadcasting system—really sixty struggling noncommercial stations—was little more than an animated corpse, a zombie that, once at rest, might not be resuscitated. The situation was saved when the BBC's *The Age of Kings*, a fifteen-part series on Shakespeare's historical plays, was imported from Britain. Frank Baxter provided an introduction for each episode, specifically tailored for American audiences.

It all began when the National Education and Radio Center (NETRC) began nursing plans to become America's fourth major network. The NETRC, soon renamed the NET (National Educational Television), felt ambitious enough to challenge the entertainment-oriented programs of the commercial networks. It wanted to connect the various nonprofit "public" stations, while at the same time lifting them out of the doldrums of pedantic banality with quality productions of real prestige.

But NETRC was basically a program distribution center with neither the license nor money to produce their own shows. They cleverly got around these obstacles by importing the BBC Shakespeare series. The timing was also right. That same year, Newton Minnow, who President Kennedy had just appointed FCC Chairman, had denounced TV as a "wasteland of westerns and detective shows." Now, public television could come forward and offer an antidote to that mindless wasteland.

But NETRC needed some kind of corporate sponsorship if it had any hope of purchasing the series from the British. There was also the question of national rights, promotion, and publicity. NETRC's Nazaret "Chick" Cherkazian and Warren Kraetzer were given the assignment to find a willing sponsor. They found one on the Humble Oil and Refining Co.—the marketer of Esso Gasoline, later Exxon.

Humble Oil President M. J. Rathbone felt this was a unique opportunity, and he grasped it. He wrote that "this was a definite step forward in providing intelligent, cultural programs on television." He was probably sincere, but there were other considerations as well. Like AT&T discovered a decade earlier, Humble Oil could present itself as a patron of the arts and erase the image of the corporation as a money-grabbing profiteer.

But there was *one* thing that Rathbone felt was absolutely essential: they must obtain the services of Frank Baxter to introduce each segment. Baxter was the foundation, the key, the essential ingredient. If they couldn't get the professor, the deal might be off. He was an insurance policy against possible disaster. Rathbone and the other Humble executives knew they were taking a big risk in money and prestige. Could the American public be weaned off westerns and mindless sitcoms?

But with Baxter on board it was a sure thing. Humble felt Baxter was, "A pedagogue of great experience" his manner "a sort of distinguished avuncularism and a gleeful excitement with his subject, and the happy opportunity of communicating it with others." As Gilbert Seldes once put it, Americans were "the kind of people who would have been sure that a college professor talking about Shakespeare was not for them; when they discovered the professor was Frank Baxter, they decided it was very much for them."

When NETRC men Kraetzer and Cherkezian went to Houston to meet with Humble Oil executives, the professor was one of the first topics of conversation. "Wouldn't it be great to get Baxter?" they pointedly enquired. Kraetzer and Cherkezian took the not-so-subtle hint, and tried to get the good doctor's services. Luckily for public television, he was available.

An Age of Kings was one of the BBC's prestige productions, and gained enormous popularity in Britain. The American "colonials" got the broadcast rights from the BBC at a bargain basement price of $250,000. The British Broadcasting Corporation agreed to this low sum in part because they wanted their programs to be better known in the United States.

Ironically, *An Age of Kings* had been shown before in the United States, and only a few months earlier. The Metropolitan Broadcasting System—much smaller than the major networks—had imported the series and shown it in a limited market—New York, Washington, and Los Angeles. It seems to have generated little comment or popularity.

In the meantime, the NETRC/public television version went forward. The whole cost of the project, including the rights, the opening and closing Baxter segments, and the publicity, was about $400,000. For the first time, NETRC and its flock of stations had enough advertising money to compete with the big networks and reach the general public. Before this, Cherkezian admitted, "we used to have about ten cents for promotion."

When Humble heard the news that Dr. Baxter was going to do it, they were elated. His role would be to make commentaries on "the historical, geographical, and genealogical backgrounds of the plays." Baxter downplayed his role, saying, "I exist only to make clear certain things that are not really part of the natural background of any normal American, something about the relationships of all these people, what struggles are in which they are involved. In a sense I am the scorecard of the game, without which you can't follow the action."

Perhaps, but he was a very important "scorecard." In fact, about twenty to twenty-five percent of the air time was given to Baxter's introductions, and a short summary he provided after each closing. He was so important that, at least in one instance, a newspaper printed his picture, not a scene from the plays, when reviewing the shows.

Baxter drove up to KQED in San Francisco, where his segments were done. In his car he had the usual array of pictures, diagrams, and charts—visual aids that he had used in thirty years of teaching. This would not be the first time in his retirement that he found himself a 300-mile-plus "commuter."

An Age of Kings began nationally on October 20, 1961, and it became an enormous hit that found favor with critics and audiences alike. The *New York Herald-Tribune* was enthusiastic, saying the series was "Easily one of the most magnificent efforts of the TV season. It was one of the nicest things that has happened to TV in years." It was incredible but true: long-suffering public television had a bona fide hit on its hands. And, while giving due credit to the BBC and its wonderful actors, much of the success in America can be attributed to one person: Frank Baxter.

After the opening credits, Dr. Baxter would appear on a stage with a few simple props. He is usually sitting in a Tudor-style chair behind a desk that holds a sword and a crown. The backdrop of the stage is a medieval banner, a draped curtain, and a shield. But there's also what looks like a miniature stage, the kind that might be used in a puppet show. Dr. Baxter walks over to the stage occasionally, pulling aside its curtains to reveal a family tree, a map, a drawing of a castle, or drawings of kings. You never know what Baxter is going to reveal, which evokes memories of the Bell Science "magic screens." At one point, there's even a contemporary portrait of Queen Elizabeth I, under whose rule Shakespeare began his career.

Dr. Baxter seems to delight in surprising his audience with little props. There's a roly-poly, round doll-like little toy with a white beard, which Baxter dubs "Falstaff." When you push the doll, it immediately rights itself, illustrating Falstaff's political balance and his ability to come up right. Or Baxter might suddenly reach in his pocket and produce an original coin of the Elizabethan period. As the camera comes close up, Baxter notes that, as we see in the coin inscription, England still claimed France as late as the 1590s. This forms a backdrop for the play *Henry V*, in which the English king campaigns in France to win its crown.

The professor often speaks in common, everyday language, sprinkled with slang. That, of course, was always part of his appeal. He says, for example, that Richard II often "steals the show." He also takes great pains to make parallels to American life and culture, so his audience can better understand as the plays unfold. Seizing the crown is a very risky and sometimes violent act, and is certainly "*not* like taking a senatorship or a governorship in an American election."

Baxter also tackles the core question for the whole series: why should Americans even *care* about historical events in the medieval past? Getting a bit poetic, the professor begins by saying "all history flows towards us out of the past like a great river and all that we have and are is conditioned by the past."

But there's more. "Without the end of feudalism (chronicled in these plays) we would have been without the growth of the… middle class, and without the importance of power of Parliament our political system would not have become possible."

Baxter pulls out all the stops by concluding:

> *Isn't it interesting to think that in Shakespeare's audience back there so many hundreds of years ago, the groundlings who paid their penny in the pit included many an Elizabethan apprentice who was to journey far across the Atlantic, live out his life, achieve some competence and success in life and leave his bones in the soil of Plymouth Colony, Massachusetts, or Virginia.*

This is typical Baxter, and well illustrates why he was such a charismatic teacher.

Frank Baxter refers to the players as British "provincial" actors—a code word for people who were not major stars. The "provincial" also refers to the fact that many players spout dialogue in North Country, Welsh, or Scots dialects, some of which are nearly unintelligible to American ears even today. Baxter and most Americans in the 1950s and 1960s were used to the more "refined" classical "Oxbridge" accents of famous actors like Sir Lawrence Olivier.

Most of these "provincial" actors were going to become very well known in years to come, though Baxter had no way of knowing the future. Julian Glover, for example, plays Edward IV. He doesn't have a "household" name, but his face is very familiar. He was the evil millionaire who is seeking immortality, and subsequently chooses the wrong cup, in *Indiana Jones and the Last Crusade*. Robert Hardy (Henry V) is another actor well known to Americans, and is noted for his portrayal of Winston Churchill.

But two of the most famous were still obscure at the time, and one of them was just starting her career. Sean Connery is "Hotspur," about a year before he achieved superstardom as James Bond in *Dr. No* (1962). His only major American appearance was in the Walt Disney leprechaun fantasy *Darby O'Gill and the Little People* (1959). Dame Judi Dench, "French Princess," won a Best Supporting Actress Academy Award for her performance as Queen Elizabeth I in *Shakespeare in Love* (1998). She's most famous, however, for her recurring role as "M" in the most recent James Bond/007 movies.

An Age of Kings was more than just a smash hit: it also established a precedent for future public television operations. To begin with, it was the first nationally distributed noncommercial series to receive support from a major commercial source—in this case Humble Oil. There would be no product commercials, but the sponsor's name would be prominently displayed as the entity that "underwrote" or gave a "grant" for the production.

Secondly, it began a public television tradition of acquiring British programs—dramas, comedies, documentaries, and light entertainment—to be broadcast here in the States. The BBC was more than willing to cooperate—"Our production, your money" was the standard saying. Within a few years, the BBC even consented to coproduce films with public stations. The British never understood, and probably still don't understand, why American public TV is always on the brink of seeming extinction. Too much time, they feel, is devoted to pleading, begging, and moaning for money with various promotions, pledge drives, and so on. In Britain, the BBC (in essence, their public TV) is funded by a TV license that you must purchase if you own a TV.

But again, without Baxter, *An Age of Kings* would not have been the success it did become. And without the BBC programming, where would PBS be today? Would there even be a PBS? It's hard to answer such a question, but in all likelihood public television would have succumbed, a victim of its lack of money and its educational roots.

The next big milestone in Baxter's busy retirement was in 1964. Writing to his mentor Dr. Colton in September of 1963, Baxter relates, "Next year is the 400th anniversary of Shakespeare's birth, and Westinghouse Broadcasting Company (for whom I have worked before) have made a tentative offer to me to become their Shakespeare advisor for all of the films and tapes to be made in connection with this celebration in 1964."

Baxter playfully adds, "They have even suggested my going to London at their expense for the month of October, but this is beyond any real possibility, I am sure. I will believe it when I find myself in the TWA passing over the southern tip of Ireland." But the idea of Baxter and the Bard's birthday was something that Westinghouse simply could not pass up. The London trip *did* become a reality, and much more besides.

A couple of months later, the professor reported back to Colton that Westinghouse did indeed hire him as Consultant for the 1964 "Shakespeare" year. His duties included "my appearances in fifty-six TV programs, a good deal of work on radio, and personal appearances of one sort or another in the seven cities covered by WBC."

Baxter goes on to say that he and his wife Lydia had just returned from a trip to San Francisco, where the Shakespeare shows were going to be taped. Taping was at Westinghouse station KPIX-Channel 5. "We drove north carrying a car load of pictures, maps, and models to be used as visuals on these programs. The weather was superb—both ways—and we enjoyed the trip immensely. I am to begin work on the actual programs in January."

But Baxter made his biggest "splash" when he travelled to London later in 1964. As the professor wrote later, "They sent me to England for a busy three weeks or so to make all sorts of film footage to be used in the series. The weather cooperated splendidly, and so did all the people with whom I had to do business."

Westinghouse "Group W" sent him back to England in April of 1964 for another round of commemorative events. Dr. Baxter stirred things up in the British papers by declaring that the Beatles' haircuts were inspired by Shakespeare's historical plays set in the middle ages. King Henry V—he of the famous lines "we few, we happy few, we band of brothers"—sported a kind of mop top hairdo, as did most of his contemporaries. Now, this was 1964, and both Britain and America rocked with the Beatles craze.

Those lads from Liverpool—John, Paul, George, and Ringo—were baffled and confused. They claimed that at one point they had been swimming and had to do a gig with their hair wet and plastered across their foreheads. They liked the look and cultivated it, though they didn't say why they grew their hair so long (by early '60s standards). But their chief question was this: "Who was that bloke Baxter?"

Hardly pausing for a breath, Baxter left the London controversy behind him and headed to Stratford-Upon-Avon for the actual commemoration ceremonies with a busload of journalists in tow. Cecil Smith, a writer for the *Los Angeles Times,* later recalled the trip vividly. The bus rattled along an English country road, and even though it was spring, it was pouring rain and the blustery wind buffeted the vehicle mercilessly,

It was lambing season, and the tiny black-faced sheep scampered along beside their sodden mothers in the green and rain-slick fields. Suddenly, the sun peeked out from behind the clouds, and a glowing beam pierced the muddy road. Just as the solar disc appeared, a lark, perhaps disturbed by the approaching bus, shot skyward.

"Stop the bus!" Baxter ordered the driver, then bolted out the door and into the muddy road. The reporters dutifully followed. "How could they say that Bacon wrote those plays?" he exclaimed. Baxter was referring to an old chestnut of a controversy, which claimed that Shakespeare was merely an actor, a front, or a façade, and that another man wrote the plays instead.

One of the candidates put forward was Francis Bacon, a philosopher and government official of the Elizabethan/Jacobean period. But Baxter was scoffing at this idea through his sudden bus stop. As the sun glistened on his bald pate, the professor stretched out his arms in a dramatic gesture. "Bacon was a city boy!" he explained. "Only a country boy would know how a lark flies—'Like a lark at break of day rising from the sodden earth…' What does a city boy know of such things, of the smell of the soil, the whisper of grass, the gamboling of lambs, the trembling of deer… Only a country boy born and bred in the country could know these things!"

The journalists climbed back into the bus, muddy but enlightened. Baxter was maybe a little damp but beaming. That lark had shot up almost on cue, and gave Baxter an opening for a wonderful little commentary. As Cecil Smith recalled from this incident, "Baxter had a wondrous gift for a teacher of using the ordinary and mundane to make his point."

About Time, Baxter's last Bell Science film, also falls within the professor's retirement years. It was filmed in 1961, but for some reason was released in 1962. Baxter himself says, in a letter, that he doesn't know why it hadn't been aired on TV. Perhaps it was a sign of Bell's growing disenchantment with Warner Brothers.

Occasionally, Baxter would make an appearance as an expert on one subject or another. On June 4, 1965, Dr. Baxter appeared on NBC's *Today Show* to be interviewed by Barbara Walters. Since this was June, and a time when hundreds of thousands of high school and college students were graduating, Baxter appeared to trace the history of the cap and gown, its evolution in twelfth century Europe, and the significance of the different color edgings. He also explained the meaning of commencement, Bachelor's Degrees, and Master's Degrees to Walters.

As usual, he used illustrations to prove his points and make the presentation more interesting. Barbara Walters wore an actual cap and gown to get into the spirit of the occasion. The segment showed that Baxter had lost none of his skills of teaching and communicating to a general audience.

In 1966, there was the documentary *Our Heritage*, which is basically a retelling of the Declaration of Independence and the birth of the United States. It was produced by the U.S. Department of Defense, which is a little odd. The Vietnam War was raging, and public opinion was increasingly divided over the conflict, a polarization that perhaps this film tried to remedy by appealing to everyone's patriotism.

After a prologue that shows a Fourth of July Parade, the film goes into the beginning credits. When Baxter is about to be introduced as host, a voiceover says, in stentorian tones, "Dr. Frank Baxter is an American. A man of many degrees and awards, including the Peabody Award in addition to seven Emmys, he is a dedicated student of American history, and proud of the one thing he loves above all things-his country. Now, let us meet our distinguished host."

Why the Department of Defense felt obligated to underscore Baxter's record and his credentials is anyone's guess. Perhaps they felt that since he was a retired English professor, not an historian, the pompous buildup was necessary. The Professor merely introduces himself as "Frank Baxter." He is on a set that includes a large copy of the Declaration of Independence, a model of Independence Hall, and a full-scale replica of the Liberty Bell.

He recounts the events leading up to the revolution, and does it with the flair and style we remember in his earlier programs. At one point the camera cuts away to show a montage of pictures illustrating points of the revolution. The images, apparently created for this film, are of poor quality, at least by today's standards. But they in no way detract from Baxter's usual brilliance.

But perhaps the best part of the twenty-eight-minute film is when Frank Baxter goes to Philadelphia and Washington D.C. and the camera tags along. He speaks of the adoption of the Declaration while standing in the inner chamber of Independence Hall. Then, he goes to Washington for what is in essence a historical tour of the capital, circa 1966. Baxter walks up the steps of the capitol, and goes inside to the rotunda just under the majestic dome. "Magnificent, isn't it?" Baxter asks.

The only moment of unintentional humor comes when he describes the famous Trumbull painting of the presentation of the Declaration to Congress. There's "Benjamin Franklin, short in stature but a giant among men," he declares. Earlier, he mentioned that "Franklin invented bifocals, you know." But the real humor is when he describes Jefferson as dressed in a "gay red vest." This was the time when gay meant happy, not homosexual!

The White House is briefly visited as well, Baxter shielding his bald pate with a typical fedora. While he isn't shown, the occupant was President Lyndon Johnson, already grappling with the pressures of the escalating Vietnam War. Only two years later, in 1968, troop levels in Vietnam would reach 540,000 men. In fact, Baxter's son, a career soldier and Korean War veteran, was in Vietnam around the period the professor was doing this show.

The film ends with a poignant visit to Arlington National Cemetery, where so many thousands of our military men lie buried. There's a brief glimpse of President Kennedy's grave and eternal flame. He had been assassinated only three years before. Not only had Baxter been a soldier himself in World War I but he must have thought about the risks his own son was taking in Vietnam. As far as can be known, *Our Heritage* is one of the last of Frank Baxter's major film projects.

By 1970, Frank Baxter's film and television career was drawing to a close. It had been a long and, in many ways, incredible run—almost twenty years. By coincidence, the *Los Angeles Times* did an interview with him that year, in which he sort of summarized his film and television career. The *Times* was mainly interested in getting his opinion on the various Shakespeare plays and films that were current at the time. There was no conscious attempt to give Baxter a platform to look at his "career" in a nostalgic and retrospective manner.

"I'm for hire," he declared, "but not for another series, please. I'm too darned old. TV hit me late in life." Smiling, he then quipped, "Now I don't know if I should go into ballet or not." The reporter wanted his opinion on Franco Zeifferelli's then current movie, *Romeo and Juliet*. At the time there was some controversy over the film, since the Italian director had cast teenagers in the lead roles. Much of the controversy revolved around Olivia Hussey as Juliet. She was beautiful and radiant, but only sixteen, and she did a brief nude scene.

"A teen-age Juliet?" Baxter asked with a frown. "I always think of a Vassar girl. You see every man makes his own image. The trouble is I've lived with these plays in my head for so long I don't like to disturb my own image."

Baxter felt the possibilities of literature on television had hardly been tapped. "I wish I had another twenty-five years to take them on," he said with a wistful smile. Still, he could look back on what he called a "very gratifying life." After all, as a teacher "I took the whole world for my province. Imagine what fun that was! And out of TV, I met people like John F. Kennedy, Joe Welsh, Hume Cronyn and Jessica Tandy."

As Baxter grew older, infirmities began to make inroads into his health, but he steadfastly refused to exercise or eat "properly." He would continue to smoke pipes or short black cigars, and when he developed lip cancer, his "solution" was to merely change the position of the pipe stem in his mouth.

His diabetes also troubled him, and the problem began to affect one leg and foot. The foot problem was probably the most bothersome, and it gave him constant pain. Pain pills gave some, though perhaps not total, relief. When the professor was in his mid-seventies, his wife Lydia wrote to Frank Junior about the situation. "Father is about the same," she reported. "The foot hurts a lot. They give him pills for the pain. The doctor says it is better to bear it than to have the foot and part of the leg off. It wouldn't do much good either way… The business of learning to walk with an artificial limb is too hard at his age."

Frank Baxter still retained a very sharp wit even into his eighties. He was honored by a star on Hollywood's Walk of Fame, but jokingly lamented it was right in front of a liquor store. He had hoped it might be outside a library! And he still held court as "Reader in Residence" at the USC campus. His annual Bovard Hall presentations were so popular that *Los Angeles Times* would print articles advertising them, with headlines like "Dr. Frank Baxter Will Visit USC." Reporters were often dispatched to interview him on or near these occasions.

Infirmities and disease notwithstanding, Baxter seems to have enjoyed his senior years. He read avidly, and though some stories say he shunned TV, that simply isn't so. His daughter, Lydia Morris Baxter, recalls, "He loved to watch *Hawaii Five-O,* with Jack Lord, and *The CBS Evening News with Walter Cronkite.*" The professor also enjoyed the Korean War comedy-drama *M*A*S*H.* This seemed to be a favorite, since, as his daughter reminds, "he was in the Medical Corps in World War I."

Daughter Lydia tried to have him take walks and exercise, but Baxter resisted her efforts. He played chess with his daughter but found it would interfere with his sleep at night. The professor would putter around in his workshop, which his family called "the Boar's Den." He liked animals, and one of the family dogs, Elizabeth, was one of his companions, along with another dog and two cats. But above all, it was reading that gave him his greatest pleasure.

One evening in January 1982, the eighty-five-year-old Baxter was enjoying a dinner with his son Frank and daughter-in-law Kaye, but as the evening wore on he just didn't look well. Kaye was a former nurse, and to her the symptoms suggested heart troubles. Though he protested vigorously, his family took him to a hospital emergency ward, where he spent his time joking and flirting with the nurses.

The doctor arrived on the scene, but they still didn't seem to know what ailed him. "Have you ever had heart problems?" the doctor asked. "Never," was the firm reply. Just moments later, Dr. Frank C. Baxter passed away from a massive heart attack. The date was January 18, 1982. He was cremated and his ashes scattered over Colorado, where his son and many family members were now living.

Baxter was gone, but his Bell Science films were still being shown. Without the Bell Science films, he might not have achieved a kind of immortality that continues to this day. His *Shakespeare on TV* shows formed a model of how academic subjects could be successfully taught on television, and there's no doubt his programs helped "save" a struggling public television system, however briefly. But by themselves, the Shakespeare shows were what one writer called an "interesting cultural artifact" but little else. Indeed, unlike the Bell films, they are forgotten today.

Three years after Baxter's passing, Jim Blinn became involved with *The Mechanical Universe: And Beyond.* It was a fifty-two-part video instructional series that helped students—high school, college, or adult—understand physics. Produced by the California Institute of Technology and Intelecom, the programs were designed to make complex subjects more accessible to average people—a concept pioneered by the two Franks: Baxter and Capra. Like James Burke's *Connections* and Sagan's *Cosmos*, these films go on location to give the programs more variety and scope.

Today, Jim Blinn says that he contributed over "eight hours of animation that I designed to show algebra, calculus, Newton's Laws, the Theory of Relativity, etc." The series also had historical reenactments of Newton and other science greats. It's very well done and very much in the Baxter-Bell mode, allowing for the inclusion of modern CGI. For this viewer, at least, the series' only slight flaw is the main lecturer, Dr. David Goodstein of Cal Tech. He might be a brilliant physicist, but his presentation harkens back to the "bad old days" of early public television. He's rather dry and matter-of-fact, without an iota of Baxter's or Burke's appeal.

By contrast, William Sanford Nye was, and is, very entertaining, though more in the mold of James Burke. In fact, during the enormously popular PBS series *Bill Nye the Science Guy*, he displayed a sort of manic energy and some fine comedic skills. He has the ability, just like his predecessors, to make science not only accessible but fun.

Nye was born on November 27, 1955, in Washington D.C. He attended Cornell University and studied mechanical engineering. Interestingly enough, one of his old professors was Carl Sagan. After he graduated, Nye started his career at Boeing—a major player in the aerospace industry. He starred in training films as a mechanical engineer, and he also developed a hydraulic pressure resonance suppressor that's still used in the 747 jumbo jet. It's been said he had the desire to be an astronaut, but he was rejected from the program.

In the early 1990s, he was featured in live action sequences of *Back to the Future: The Animated Series*. He didn't have any lines but silently demonstrated the experiments that "Dr. Emmett Brown" (actor Christopher Lloyd) described in detail. That exposure led to the *Bill Nye The Science Guy* series. Since it was intended for a pre-teen audience, Nye's persona is a bit more manic and over the top, but always entertaining and full of solid facts.

Bill Nye's *Science Guy* always wore a blue lab coat and bow tie, making it kind of a signature costume for the series. The topics covered were wide ranging: everything from dinosaurs to human skin to volcanoes. In the hurricane episode, Nye is riding a bicycle, chased by a violent storm. "A twister! A twister!" he cries, frantically pedaling. School children would probably miss the oblique homage, but the segment has elements of the famous Judy Garland movie *The Wizard of Oz*.

Bill Nye the Science Guy ended in 1997 after 100 episodes, but Nye's alter ego still appears from time to time. The Epcot segment of the Walt Disney World in Florida features a video entitled *Ellen's Energy Adventure*, in which "Science Guy" costars with comedian Ellen DeGeneres and quizmaster Alex Trebek. Nye also lent his voice talents in the "Dinosaur" attraction in Disney's Animal Kingdom Park.

In 2005, he created a thirteen-episode PBS KCTS series called *The Eyes of Nye*, a science show that was aimed for an adult audience. Nye is also a fellow of the Committee for Skeptical Inquiry—an organization based on promoting science and the use of reason. In fact, in 2011, he received an "In Praise of Reason" award from the Committee of Skeptical Inquiry for his tireless efforts in education.

Bill Nye has been a frequent guest on CNN and other programs, particularly on the *Larry King Show.* A debunker of UFO stories, he had a sparring match with a very irate Dr. Bob Jacobs on the matter. In the exchange, Nye showed himself to be more in command of his emotions, and he was a gentleman at all times. Nye has also appeared on various shows to discuss such timely topics as global warming. In 2014, he remains one of the most articulate and well known "faces" of popular science.

Though considerably modernized, the Baxter-Bell method of science popularization is alive and well in the twenty-first century. A shining example of a modernized "Baxter-Bell formula" series would be the Science Channel's *Through the Wormhole*. It began airing in June, 2010, and continues as of this writing. It combines many of the techniques that had gone on before, including stunning CGI (as in *Cosmos*), scenes involving actors (as in *Connections*), and filmed out-of-the-studio segments like the Bell films. There are several links to the Bell Science shows, including its optimistic view of the world, its stimulating curiosity about "nature, and nature's laws," and above all, a charismatic host/narrator.

Academy Award-winning actor Morgan Freeman is the host of *Through the Wormhole,* and he shares Frank Baxter's enormous audience appeal. "All of these (*Wormhole*) topics fascinate the heck out of me," Freeman says on a *Science* website, and it shows. He has all the infectious enthusiasm about learning that Baxter so wonderfully displayed.

Through the Wormhole can claim "descent," in a way, from the Bell Science series and the other shows that followed it. And there is a connection to the scientific community that echoes the Bell films, though the relationship is less rocky than what Capra had experienced. *Wormhole* producer James Younger said, "We have a strong relationship with real scientists, academic professors who are experts in math… physics, cosmology, biology."

The series tackles some very interesting topics, like time travel, quantum mechanics, the possibility of life on other planets, and whether there is a sixth sense. In one way, *Wormhole* is unlike Bell films: it dares to tackle subjects that would be taboo—or anathema—to devout believers like Frank Capra. One would hardly imagine Capra producing an episode like the one that started the *Through the Wormhole* series. Entitled "Is There a Creator," the show provocatively asks if the universe was just a coincidence, or was it created by a God that looks after us.

It's been sixty years since Dr. Frank C. Baxter first became an education superstar, and almost sixty years since the first Bell Science feature, *Our Mr. Sun,* first aired on television. Baxter is a key figure in both the development of the popular science film, and the "rescue" of the infant that would later become The Public Broadcasting System. His is a rich legacy.

Bibliography

Alexander, Goeff. *Academic Films for the Classroom: A History.* Jefferson, NC: McFarland, 2010.

Brown, David R. *Shakespeare for Everyone to Enjoy* Daly City, CA: Dry Ink Publications, 2007.

Burke, James. *Connections.* Boston: Little, Brown and Company, 1978

Burns, Eric. *Invasion of the Mind Snatchers: Television's Conquest of America in the Fifties,* Philadelphia, PA: Temple University Press, 2010.

Capra, Frank. *The Name Above the Title.* New York: Macmillan, 1971.

Dent, Marjorie, ed. "Frank Condie Baxter" *Current Biography 1955.*

Gilbert, Jack. *Redeeming Culture; American Religion in the Age of Science* Chicago: Chicago: University of Chicago Press, 1997.

Hawes, William. *Public Television: America's First Station* Santa Fe, NM: Sunstone, 1996.

LaFollette, Marcel Chotkowski. *Science on the Air: Popularizers and Personalities on Radio and Early Television* Chicago: University of Chicago Press, 2008.

________. *Science on American Television: A History* Chicago: University of Chicago Press, 2013.

Lennon, Patricia. "*The Age of Kings* and the Normal American" *Shakespeare Survey 61* edited by Peter Holland, 181-198. Cambridge: Cambridge University Press, 2008.

Maltin, Leonard. *The Disney Films,* 3rd edition NY: Hyperion, 1995.

Manchester, William. *The Glory and the Dream: A Narrative History of America, 1932-1972,* Little, Brown and Company 1974.

Miller, Jimmy. *The Life of Harold Sellers Colton* Tisale, Arizona: Dine College Press, 1991.

McBride, Joseph. *Frank Capra: The Catastrophe of Success* NY: Simon and Schuster, 1992.

Mergen, Bernard. *Weather Matters: An American Cultural History Since 1900.*Lawrence, KS: The University Press of Kansas, 2008.

Poague, Leland. *Frank Capra: Interviews.* Oxford, MS: University Press of Mississippi, 2004.

Sagan, Carl. *Cosmos* NY: Random House, 1980.

Scramm, Wilbur, Jack Lyle, and Ithiel De Sola Pool. *The People Look at Educational Television,* Stanford: Stanford University Press, 1963.

Sito, Tom. *Moving Innovation: A History of Computer Animation.* Boston: Massachusetts Institute of Technology, 2013.

Smith, Roger P. *The Other Face of Public Television: Censoring the American Dream* NY: Algora Press, 2002.

Smooden, Eric. *Regarding Frank Capra: Audience, Celebrity, and American Film Studies, 1930-1960.* Durham, North Carolina: Duke University Press, 2004.

Stewart, David. *The PBS Companion: A History of Public Television* TV Books, 1999.

Thomas, Gerald W. and Donald G Ferguson. *In Celebration of the Teacher* Las Cruces, NM: New Mexico State University, 2013.

MAGAZINES AND NEWSPAPERS

Adams, Val. "Now and Then: Dr. Baxter Talks About his Discussion Series," *New York Times,* Aug 8, 1954.

Ames, Walter. "Professor at SC Wins TV Honors" *Los Angeles Times* Feb 12, 1954.

Cuno, John Marshall. "Discovering the Sea" *Christian Science Monitor,* Jan 25, 1964.

____________. "Sojourn in Old London" *Christian Science Monitor* June 12, 1964.

Dornbrook, Don. "Ernie Ford Has a Word for It" *Milwaukee Journal* July 14, 1957.

"Dr. Frank Baxter Assails 'Tropic" and calls It Obscene" *Los Angeles Times,* Feb 9, 1962.

"Dr. Frank Baxter Will Visit USC" *Los Angeles Times,* Dec 12, 1965.

Drake, Hal. "Baxter Will Retire at Semester's End" *Daily Trojan* No. 122, May 10, 1961.

Gould, Jack. "Television: Our Mr Sun, Bell Telephone System Offers Science Program at Prime Time" *New York Times,* Nov 20, 1956.

___________. "TV: The Story of Blood 'Hemo the Magnificent' Is a Fascinating Account of Circulation and the Heart" *New York Times,* Mar 21, 1957.

__________. "TV: A Study of Genetics: Program Marks the Return of Bell System's Science Series, With Dr. Frank Baxter" *New York Times,* Dec 10, 1960.

Handy, Mary. "Teaching on TV is an Art, Says Baxter" *Christian Science Monitor,* Jan 19, 1957.

Hartford, Margaret. "Shakespeare's TV Friend for Hire, but Not in Series" *Los Angeles Times,* Jan 2, 1970.

Jones, William M, and Andrew Walworth. "Saudek's Omnibus: Ambitious forerunner of Public TV" *Current,* Dec 13, 1999.

Lara, Adele. "KQED at 50: KQED is an Institution in Public TV, but from the Beginning It Took a Anything-Goes Approach" *San Francisco Chronicle,* Apr 28, 2004.

Life Magazine, "TV and Teachers Team Up: Professor's like California's Erudite and Witty Baxter Show the Way to Video's Instructional Role" Dec 7, 1953.

Lynch, Samuel Dutton. "Dr. Frank C. Baxter, Shakespeare, and TV" *New York Times,* Feb 21, 1954.

MacCann, Richard Dyer. "TV Teacher Explores Literary Past" *Christian Science Monitor,* Dec 22, 1959.

McCoster, John E. ""Earl Herald" *California Wild,* Spring 2003.

Niderost, Eric. "Prof Frank Baxter, AKA Mr. Research" *Filmfax #124,* Jan 2010.

__________. "From Sonnets to Sunspots: The Career of Frank Baxter, Television's First Academic Superstar" *Nostalgia Digest* Autumn 2012.

__________. "Dr. Frank Baxter—You May Remember Him as 'Dr. Research'" *Looking Back,* Feb/Mar 2012.

Rummel, Frances V. “TV’s Most Surprising Success” *Reader’s Digest* Sept 1956.

Smith, Cecil. “Frank Baxter: He Gave Meaning to the Gogglebox” *Los Angeles Times,* Jan 31, 1982.

Smith, Jack. ”Dr. Frank Baxter Delivers Final Lecture at SC, Makes Sparks Fly” *Los Angeles Times* May 10, 1961.

________. “Baxter Loved from Ghost to Ghost” *Los Angeles Times,* Feb 3, 1982.

Solomon, Charles. “Science Films of ‘50s Not Just a Memory Anymore” *Los Angeles Times,* Oct 13, 2003.

Stewart, David. “Frank Baxter, Television’s First Man of Learning” *Current,* Jan 29, 1996.

____________. “An Age of Kings: an Import Becomes Public TV’s First Hit” *Current,* Jan 21, 1998.

Templeton, David. “Weird Science” *Sonoma County Independent* Sept 23-29, 1999.

“TV Prof Makes His Own Props” *Popular Science,* Mar 1955.

Warner, Jack Jr. “The Story of a Lost Film” *The Keaton Chronicle* Vol. 1, Issue 4, Autumn 1993.

PRIMARY SOURCE MATERIALS; CORRESPONDENCE

Dr. Frank C. Baxter, Letters to Dr. Harold Colton, 1926-1967, Manuscript Collection, Museum of Northern Arizona.

________________, Letters to Frank Capra, 1956 and 1961. The Frank Capra Cinema Archives, Wesleyan University.

Lydia Foulk Spenser Morris Baxter (Dr. Frank Baxter's wife) undated letters to son Frank Baxter "jr"—courtesy Allison Blackburn, daughter-in-law of Dr. Baxter.

UNIVERSITY OF SOUTHERN CALIFORNIA YEARBOOKS

El Rodeo Yearbook—Class of 1939.

El Rodeo Yearbook—Class of 1952.

El Rodeo Yearbook—Class of 1956.

El Rodeo Yearbook—Class of 1957.

El Rodeo Yearbook—Class of 1958.

El Rodeo Yearbook—Class of 1959.

ONLINE SOURCES

James Burke Channelle, a collection of videos from the *Connections* series, www.youtube.com/usu/JamesBurkeWeb.

Dr. R.W. Donnell Blog, "Meteora, The Unchained Goddess", DoctorRW.blogspot.com/2009/02/meteora-unchained-goddess.1958.html.

Houston PBS History, www.houstonpbs,org/aboutus/history.

S.W. Frauenfelder, Breakfast with Pandora, Nov 30, 2007 mythtypepad.com/2007/meteora-the-unc.hmtl.

Dr. Amber Jenkins, NASA climate change blog, "My Big Fat Planet," clamate.nasa.gov/blog/447.

Marcel LaFollette, A Survey of Science Content in U.S. Television Broadcasting 1940s-1950s: The Exploratory Years.

Michael Sporn Animation Blog, "Our Mr. Sun" www.Michaelspornanimation.com.

White Wilderness, www.snopes.com/Disney/film/lemmings.asp.

Index

CPSIA information can be obtained at www.ICGtesting.com
Printed in the USA
LVOW01s0309220215

427853LV00002B/17/P